Solving
The Price Is Right

How Mathematics Can Improve
Your Decisions on and off the Set of
America's Celebrated Game Show

Justin L. Bergner

Prometheus Books
Guilford, Connecticut

Prometheus Books

An imprint of Globe Pequot, the trade division of The Rowman & Littlefield Publishing Group, Inc.
4501 Forbes Blvd., Ste. 200
Lanham, MD 20706
www.rowman.com

Distributed by NATIONAL BOOK NETWORK

British Library Cataloguing in Publication Information Available

Library of Congress Cataloging-in-Publication Data Available

ISBN 978-1-63388-851-7 (cloth) | ISBN 978-1-63388-852-4 (ebook)

∞™ The paper used in this publication meets the minimum requirements of American National Standard for Information Sciences—Permanence of Paper for Printed Library Materials, ANSI/NISO Z39.48-1992

For my wife, Lisa,
who believed in my vision to turn my passion for mathematical games into a book.
And to my boys, Matthew and Alex,
who learned about the power of hard work to fulfill your dreams.

Contents

vi   *Contents*

Acknowledgments

I owe an enormous debt of gratitude to those individuals who helped make *Solving "The Price Is Right"* possible. Writing this book has been a lifelong goal starting over twenty years ago, and the help of these individuals was instrumental in completing this journey.

I want to first acknowledge the invaluable help of Bowman Dickson, upper school mathematics teacher at St. Albans School in Washington, DC, my alma mater. Bowman served as my primary math reviewer. Bowman was everything one could want in a math editor, not just catching the occasional inconsistency or computational error but helping me present material as a math teacher would to a class of students. Bowman even doubled as a copy editor when I needed to tighten up the manuscript. I also enjoyed working with Adam Persing. With a PhD in statistics, Adam provided a second mathematical review, bringing a fine attention to detail at a point in the process where a fine attention to detail was needed.

I also want to thank Chris Avery, professor of public policy at Harvard's Kennedy School, who advised me on a game-theory-focused senior essay in 1998 when he was a visiting professor at Yale. I reconnected with Chris Avery as I was putting the finishing touches on my manuscript. An applied game theorist and author in his own right, Chris took an interest in my discussion of Contestant's Row bidding, providing useful suggestions on how to integrate game theoretic conclusions with behavioral remedies to articulate a cohesive bidding strategy.

I owe a special thanks to my wife, Lisa. She was a very positive influence during this multiyear effort. A trained accountant, she made sure that the numbers in the text tied with the numbers in the tables so readers could more effectively follow my reasoning.

I could not have reached this point without the assistance of John Willig, my talented literary agent. John has a special ability to coach writers focused on research-based nonfiction. He grasped not only the potential of my research and manuscript but how to leverage my background in business and finance to provide real-world analogues for a number of the learnings from the show. John also made the literary process enjoyable for me, a first-time author.

Finally, a special thanks to Jake Bonar and the team at Prometheus Books and Globe Pequot, who helped me realize my vision for this book. I am thrilled that Solving "The Price Is Right" can join Prometheus's growing collection of books that popularize math for the broader public. Jake did a wonderful job managing a diverse set of graphics, by no means an easy task, which helped bring Solving "The Price Is Right" to life.

Hopefully The Price Is Right and this book will live on for decades, if not generations to come.

Introduction

About ten years ago I was at the Stamford, Connecticut, courthouse for jury duty. A large group of people—I would estimate about seventy-five—waited in a large room while the process of separating those who would be released without screening from those who would be further screened for jury duty ran its course. Most people were occupied with books or their smartphones. There was a decent-sized TV at the front of the room, which at the 11:00 a.m. hour was tuned to *The Price Is Right*. Lo and behold, the first contestant to get on stage was an energetic college student, and she was playing the pricing game *Lucky Seven* for a car. In unison, most of the seventy-five people in the courthouse stopped what they were doing to turn their attention to the show. The college student won the car in dramatic fashion, but what was far more dramatic was the collective attention diverted to the show. Young, old, male, female—it didn't matter. For a few minutes *The Price Is Right* was the only game in town for those awaiting jury duty at the Stamford courthouse. Why? Because we all grew up with the show. And because many of those individuals were surely playing *Lucky Seven* in their heads that day.

I grew up watching *The Price Is Right (TPIR)*. During the summer and on snow days, I tried not to miss it. There were so many exciting aspects to the show, between bidding in Contestant's Row, the Showcase Showdown, and dozens of exciting pricing games. The contestants were interesting characters, and Bob Barker and the Barker beauties added mystique to the show.

As a game show, *TPIR* offers a stimulating combination of knowledge, strategy, and luck. I have always found this combination to be the hallmark of a well-designed game, whether a card game, a board game, or a game show. I think this is a key reason the show endures, having completed fifty consecutive seasons in its current form. It is the second-longest continuous-running series on TV, only surpassed by CBS's *60 Minutes*, which just completed its fifty-fourth season.

The Price Is Right has outlasted most game shows that launched later. The morning primetime TV calendar no longer includes game shows like *Press Your Luck*, *Wheel of Fortune*, or *Let's Make a Deal*. *Family Feud* and *Wheel of Fortune* have persisted with some interruptions, albeit not in an a.m. timeslot. *Let's Make a Deal* has come and gone multiple times, while *Press Your Luck* was in the game show graveyard until a recent primetime revival. Each of these other game shows, while entertaining, lacks a meaningful strategy component. The Monty Hall problem in *Let's Make a Deal* is one element of strategy—but only one. And strategy in *Wheel of Fortune*—namely picking the most common consonants and vowels in the alphabet first—is fairly superficial. In fifty years, no studio has created another game show that is so rich in strategy and that combines knowledge, strategy, and luck so effectively.

The Price Is Right, by contrast, is a strategic and mathematical wonderland. Viewers have always sensed the math behind the show and its various components, even if they have not fully understood it. They know that probabilities dictate the value at which one should stop after his first spin in the Showcase Showdown, even if unable to calculate that number. Viewers know there is strategy to bidding in Contestant's Row. Most realize that the last bidder should bid $1 above one of the prior three bids or bid $1 if he strongly believes the other three contestants have gone over. But how should the first three bidders bid, knowing that the last bidder, if sensible, will pursue the aforementioned strategy?

I have wanted to write *Solving "The Price Is Right"* for more than a decade. CBS All Access helped make this possible, giving me easy access to 189 shows from season 47, which ran from September 2018 to July 2019, and 167 shows in season 48, which ran from September 2019 to June 2020.

When I set out, the goal of this book was to introduce *The Price Is Right* to the mathematical world and to introduce the show's more than five million average daily viewers to all the exciting math behind the show. To create a reference manual for regular viewers, a book of interesting problems for educators, and an introduction to the world of pop culture math for analytically minded individuals.

As I wrote the book, colleagues asked me what I thought were the grand conclusions. I initially did not have a great answer, as each game is its own mathematical problem with its own solution. I set out to write a book that compared contestant performance to optimal strategic play. I had expected performance to reflect a watered-down version of those strategies, sort of like an error-prone computer. What I found was better-than-expected rationality and strategy in certain areas, but also a set of biases and phobias in other areas that limited strategic play far more than I anticipated. Contestant behavior is where the general conclusions in the book lie, with numerous prescriptions for our everyday lives.

In *Solving "The Price Is Right,"* I have attempted to reverse engineer most of the math behind the show. Not all the math was straightforward, and I persevered through some difficult problems. But my achievement should not overshadow the mathematical talent that created the show's games and fine-tuned those games to produce the desired winning frequencies and expected win values. And they modified the games over the years to improve them mathematically and aesthetically while preserving their classic, timeless appeal.

This book is dedicated to my wife, Lisa, and my boys, Matthew and Alex. They put up with hundreds of "Come on downs" and *"ding dings"* during my two years of binge-watching the show. My boys each watched more than a hundred episodes, becoming more curious intellectually and mathematically along the way.

I began my foray into the world of advanced math as a student of Peter Kelley, now academic dean of St. Albans School in Washington, DC. With dry wit, Peter introduced me to the eye-opening world of calculus, which was an entrée to the extraordinary world of advanced math. Learning about limits and using them to calculate derivatives under Peter's teaching will always be one of the most illuminating academic exercises of my life.

After high school I completed a combined major in economics and mathematics at Yale University. In economics, I focused on game theory. My senior thesis explored the "roommate problem"—namely how to create a bidding system that divvies up rental rooms efficiently based on roommates' preferences and ability to pay while raising the necessary total rent. I was advised by Yale University James Tobin Economics Professor John Geanakoplos and visiting Harvard Kennedy School professor Chris Avery. I also enjoyed the mentorship of Stanford Graduate School of Business professor and game theorist Jeremy Bulow, who taught a seminar in "auctions and pricing" as a visiting professor at Yale.

In mathematics, I enjoyed courses in probability theory. Probability has been integral to how I think and see the world, and it is a blessing to be able to apply it so broadly in this book. *Solving "The Price Is Right"* does not require an advanced math degree. Some of the probability and game theory may require college math coursework for a fuller understanding. If you lack these tools, it is not too late to learn them. Probability and game theory are eye-opening topics in math, worthy of study into one's adult years.

Solving "The Price Is Right" focuses on the strategic, heuristic, and psychological methods that can improve contestant performance on the show. Heuristics refer to pricing conventions that could change at any time, like car prices often ending in zero in certain pricing games. Strategy refers to "game plans," usually mathematically underpinned, that will improve a contestant's outcome regardless of pricing conventions. The strategies are inherent in the design of the games, whether in Contestant's Row, the Showcase Showdown, or the show's seventy-seven pricing games. Psychological adjustments speak to undoing the biases that can limit contestant success on and off the show.

All of the observations in this book, except where noted, are my own, based on the full data set across seasons 47 and 48. I have avoided discussing product knowledge as it relates to the pricing of smaller items and larger prizes. Clearly product knowledge is another key to success on the show. But it was of less interest to me and is the subject of the 2017 documentary, *Perfect Bid: The Contestant Who Knew Too Much*.

I am thrilled to share the mathematical underpinnings of *The Price Is Right* with the show's millions of viewers. The show has maintained its mathematical mystique during Bob Barker's thirty-six years as host (1972–2007) and Drew Carey's fifteen-year-plus tenure (2007–present). So buckle up! Put your math caps on! Let's get ready to reverse engineer America's favorite game show.

1

Key Observations

Given the diverse construct of games across *The Price Is Right (TPIR)*, there is no universal strategy or set of heuristics to highlight, mathematical or qualitative. Each game is very different. What stood out were the contestants' psychological biases and anxieties that limited strategic play. Some of the biases are likely exploited by *TPIR* for dramatic effect or to manage the win rate in certain pricing games. Although I am not an expert in psychology, the biases are clear, and one can hypothesize the likely causes.

To understand the biases, imagine yourself as a contestant on the show. When you are told in Contestant's Row that the winning bid is the bid that is closest to the retail price without going over, it is easy to fixate on the "not going over" condition of winning rather than the need to be closer to the retail price than the other three bidders. If you are playing *Range Game*, would *TPIR* really make you wait thirty-five seconds for the range finder to fall into the middle of the $600 range, where most prize values fall? Most of these psychological biases are readily correctable. Simply remind yourself that there is no benefit in making the second-best bid in Contestant's Row—and that *TPIR* includes the full $600 range in *Range Game* for a reason. You may be surprised that the biases observed on the show appear in numerous other walks in life, particularly in business and finance.

SEVEN PSYCHOLOGICAL SUPPOSITIONS

Note the term "he" and "his" are used throughout this book to indicate both men and women.

1. *Underbidding bias*: Underbidding in Contestant's Row was egregious. The highest bid won a staggering 53% of the time. The proclivity to underbid by Bidders 1–3 was so large that the last bidder won a jaw-dropping 64% of the time when he made the highest bid. Underbidding was also evident in many pricing games and in the first showcase.
2. *Anchoring in Contestant's Row*: There are many examples of the blind following the blind in Contestant's Row bidding. Anchoring was most notable for low initial bids and high initial bids, with the later bidders often following with low or high bids. Bidder 2 and Bidder 3's influenced bids had the effect of narrowing the range of possible prize values over which their bids would prevail.
3. *Showcase second-bidder paranoia*: The contestant bidding on the second showcase played it too tight with his bids and overbid so much as to reduce his overall odds of victory. Overbidding was particularly pronounced for lower priced showcases presenting second, occurring 45% of the time.
4. *Good innate probabilistic thinking*: Decisions about when to spin again and when not to spin again in the Showcase Showdown were mostly on target and are an indication of our good intuitive grasp of probabilities. The same can be said about decisions to stop for cash rather than play for the main prize in certain pricing games.
5. *Poor out-of-the-box thinking*: The pricing games *Cover Up* and *Grocery Game* test a player's out-of-the-box probabilistic thinking, and most contestants failed on this front.
6. *Overbidding with reference price*: Contestants chose high-priced options too frequently in pricing games with reference prices, including *Double Prices* and *Coming or Going*. In contrast, underbidding was pronounced when there was no reference price, for example, in *Check Game*.
7. *Time is not your enemy*: Contestants underachieved in games in which they needed to wait for the price to get high enough, including *Card Game*, *Range Game*, and *That's Too Much*. Contestants were not so much impatient as they were anxious about not acting soon enough to call the price.

Let's discuss each of these shortcomings, possible causes, and applicability outside of the show.

UNDERBIDDING BIAS

If I could suggest one adjustment to improve contestant performance on the show, it would be to do away with the underbidding bias in Contestant's Row, which serves to further advantage the last bidder.

Reasons for potential underbidding range from the fear of going over to the unfamiliarity of the item up for bids. Being accustomed to sale prices and being unaccustomed to high-end brands or features also led to underbidding. Why the fear of going over? Winning in Contestant's Row is presented as "closest to the actual retail price without going over." *Contestants' first reaction is to think that they cannot win if they go over, rather than to think that they have little chance of winning if they do not risk going over.* The initial conditioning makes them concerned about going over, which causes them to underbid.

How do I know that underbidding disadvantaged earlier bidders? Bidder 4, who won 40.6% of bidding rounds, won 64% of the time with the highest bid and could have won 50% percent of bidding contests had he always bid $1 more than the prior highest bid. Moreover, overbids only represented 24% of bids. However, most situations where all four contestants overbid were unaired, and it is likely that overbids represented closer to 30% of all bids. Overbids should represent about 40% of bids with strategic play and be more prevalent among earlier bidders.

Interestingly, in the sixth and final bidding round, contestants overbid with a higher 31% frequency. By that point, contestants fear not making it out of Contestant's Row. They snap to their senses and underbid less meaningfully. The conclusion here may be that we are quite attuned to whatever we are presently nervous about, and that can throw off our decision making. Look no further than the COVID-19 pandemic. In spring 2020, I saw folks walking on the shoulders of busy streets to avoid passing someone on the sidewalk, trading a very small risk of contracting COVID-19 outdoors with a much higher risk of being hit by a car.

Unfamiliarity of the item up for bids is also a key factor in underbidding. When people do not understand an item's value, they value it less. Whether shopping at the store or buying electronics, they are less likely to pay up for a product they do not understand. In the business world, when companies are being sold, bidders often increase their bids only after they are provided with additional information with which to value the businesses. Clearly bidding in Contestant's Row is not shopping in a store or buying a company. One has to be closest to the actual retail price, versus making a bid tied to features of the product that the contestant understands.

BID ANCHORING IN CONTESTANT'S ROW

The goal in bidding in Contestant's Row, as I detail in chapters 2 through 5, is to put forth a bid that can win over a wide range of likely prize values. Specifically, how can a contestant's bid, when stacked up versus prior bids and bids yet to come, maximize his probability of winning the item up for bids? Bidder 1's bid should *strategically* influence Bidder 2's bid, and Bidder 2's bid should *strategically* influence Bidder 3's bid. Bidders should want to place bids that have a good probability of prevailing, rather than offering the best point estimate. Only the last bidder has complete control over which prize values his bid will prevail, and the next-to-last bidder also has a good degree of control. Earlier bidders need to consider how Bidder 4 will react to their bids, lest Bidder 4 essentially zero out the range of prize values over which they can win by bidding $1 more.

Notably, Bidders 2–4 often anchor their bids to prior bids. If Bidder 1 bids low, Bidders 2 and 3 are more likely to presume that the price of the item up for bids is low and bid accordingly. If Bidder 1 bids high, Bidders 2 and 3 are more likely to bid high. Even if Bidders 2 and 3 do not materially adjust their estimates, they may subconsciously deliver bids that are anchored by prior low bids or high bids. To some extent a bidder should be influenced by prior bids, but not to such a large extent that it leads to the "blind following the blind." Moreover, his attention may then drift toward trying to deliver the most accurate bid versus the most strategic bid—namely, one that is likely to secure a portion of the number line offering the greatest probability of success.

Evidence of outsized anchoring in Contestant's Row is visible in the larger-than-expected number of rounds with zero overbids and the larger-than-expected number of rounds with three bidders going over. Why? When Bidder 1 goes low, it is very common for subsequent bidders to also go low, creating a zero-overbid outcome. Likewise, when Bidder 1 bids very high and overbids, Bidder 2 and Bidder 3 often follow Bidder 1 in bidding too high. It is impossible to know with certainty what are rational versus irrational adjustments to prior bids on *The Price Is Right*. But it is noteworthy how often impressionability, rather than strategy, drives the behavior of Bidders 2 and 3, who should be focused on securing a sizable portion of the number line.

Readers should not be surprised about bidder anchoring on *TPIR*. Bidders in auctions and companies bidding on assets are influenced by other bidders, although usually in the direction of paying too much. When investing, individuals often anchor to their purchase price as a gauge of when to sell a stock, frequently holding on to an underwater investment until it returns to their purchase price. Anchoring can affect a broad swath of market participants as well, causing a stock's trading multiple to remain at a discount or premium to peers based on prior business conditions rather than the current outlook.

SHOWCASE SECOND-BIDDER PARANOIA

Excluding special shows with higher showcase values, the contestant bidding on the second showcase went over 30% of the time, versus 19% on the first showcase. To put this in perspective, 30% is likely higher than the percentage of overbids in Contestant's Row, despite three competing bids in Contestant's Row versus one in the showcase. I show how bidders in Contestant's Row do not bid aggressively enough, but also how contestants bidding on the second showcase bid too aggressively. Because the contestant bidding on the second showcase went over 30% of the time (24% of the time, excluding double overbids), he would need to have won two-thirds of the time he did not go over to match the success of the first bidder. Clearly this is not a winning formula.

Nor is the tendency to go over on the second showcase tied to the magnitude of the bid on the first showcase. Often the contestant bidding on the second showcase simply got nervous after witnessing the other player bid on the first showcase. He assumed the first bid was competitive, causing him to bid too aggressively, for no other reason than that the first bid was made. This dynamic is pronounced for less expensive second showcases. For more expensive second showcases, the anxiety also drove more aggressive bids, but this anxiety favorably offset a proclivity to underbid on such showcases.

What does this mean for human behavior? Consider a head-to-head sequential contest in which the first contestant has played but not been judged—for example, the closing arguments in a court case. Might the defense respond too forcefully after hearing the prosecution deliver its closing argument, squandering what should be a benefit in going second? Or, in the final round of a job interview, might a candidate get nervous about the other one or two interviewees and try to oversell himself?

GOOD INNATE PROBABILISTIC THINKING

The first and second spinners in the Showcase Showdown generally made sound decisions as to when to spin again and when not to spin again. The first spinner should spin again at $0.65 and less. For the most part he did, only stopping occasionally at $0.65 or less. Same for the second spinner, who should spin again at $0.50 or less, provided he has exceeded the total of the first spinner. For the most part he too made the right decision.

Sound probabilistic decisions can be witnessed in *Let 'Em Roll, Pass the Buck, Spelling Bee,* and *Temptation,* with the contestant correctly choosing when to stop for cash or prizes versus playing on for the car. Contestants understand probabilities, even if they cannot make perfect calculations on the fly in their heads. This should come as no surprise. Probabilistic thinking is an evolutionary selected trait.

BAD OUT-OF-THE-BOX PROBABILISTIC THINKING

Whereas contestants showed sound probabilistic thinking in straightforward games, they mostly failed to apply out-of-the-box strategies to improve their odds in more complex games. Consider the pricing game *Grocery Game*, in which contestants buy a self-specified number of different grocery products such that the total value falls between $20 and $22. Many tried to win with a specified quantity of one item, squandering the optionality and superior odds associated with choosing various quantities of multiple items to gradually build up to the $20 to $22 level. A similar dynamic occurred in the pricing game *Cover Up*, which we consider at length in our discussion of car pricing games in chapter 18.

It should come as no surprise that *TPIR* contestants predominantly fail at out-of-the-box thinking in an unfamiliar probabilistic game carried out in a time-pressured situation. People struggle with out-of-the-box solutions in familiar, non-probabilistic situations, before considering probabilities and time constraints. Still, *Cover Up* and *Grocery Game* can each be made easier by extending the game. Whether in a job interview or the negotiation of a government fiscal bill, there are benefits to extending the game to work through trickier issues.

ERRING HIGH WITH REFERENCE PRICE

It might surprise readers to learn that contestants are biased toward choosing too high of a price in certain pricing games. Overbidding generally occurs when the contestant has a reference price and is given two options, including in *Double Prices*, *Coming or Going*, and *Side by Side*. Similar to overbidding on the second showcase, this behavior runs counter to the more frequent bias to underbid in Contestant's Row and in pricing games without reference prices.

How does one explain this dichotomy if one believes a contestant's natural inclination in *The Price Is Right* is to underbid? Contestants seem to *not* want to believe the show is stingy, which it is not. But *TPIR* creates a range of prizes and prize values, some less valuable than others. About 30% of prizes in pricing games are worth less than $8,000. When a contestant sees two possible prices for a prize, such as $6,650 and $8,750 in *Side by Side*, he leans toward $8,750. Contestants may anticipate more potential regret if they err by guessing the less expensive price, as they generally believe that *TPIR* does not want to give away a more modest prize. We see an analogue between luxury goods and contestants erring high. For certain luxury goods, the price conveys quality and prestige, and the consumer becomes more likely to purchase the item in question if the price is higher (to a point).

Conversely, when there is no reference price, such as in *Check Game* and *Magic Number*, the contestant is less likely to see the underlying value in the prize and therefore underbid. In this setup he may be less inclined to associate a lower priced prize with *TPIR* being stingy.

TIME IS NOT YOUR ENEMY

Contestants underachieved in all three pricing games when it took time for the price to increase to the target level or range, notably *Card Game*, *Range Game*, and *That's Too Much*. Waiting for the price to get high enough seemed to make contestants anxious, and that anxiety frequently caused them to act too soon. (Note that the price moves in only one direction in these games.) *Card Game* is the best example, as it takes an average of 8.5 draws to get from the $15,000 starting point to a $21,000 price, assuming no ace draw. No car in *Card Game* was ever priced less than $21,000.

The real-life lesson is that we are not conditioned to succeed at games that make us wait. Knowing that we need to make a decision while the clock ticks makes us anxious, and that anxiety can cause us to act too soon. Consider buying a house. The best way to get a good deal on a house purchase, if not in a super tight market and if not in a rush to move, is to wait for the list price to be lowered. If worried about the house being sold to another buyer, inform the selling realtor of your interest and ask that he contact you if there is other buyer interest. But waiting causes anxiety, which often causes buyers to make an offer before the price is lowered.

Part I

Competitive Contests: Performance, Strategy, and Life Lessons

2

Contestant's Row Observations

Come on down! You are the next contestant on *The Price Is Right*! Bidding in Contestant's Row is the epicenter of the show. It occurs six times on every show, takes up close to one third of the show's airtime, and must have a winner each round. Getting called down to Contestant's Row from the audience is the height of excitement for an attendee of the show. Only nine out of roughly three hundred individuals in the audience, or 3%, are called down to Contestant's Row. They are selected based on their enthusiasm, eccentricities, and with an eye toward diversity across the show.

Six of the nine participants in Contestant's Row make it on stage to play a pricing game. About half of the six win their pricing game. Two of the six who get out of Contestant's Row win the Showcase Showdown to make it to the Showcase, where one contestant wins a fabulous collection of prizes typically valued between $20,000 and $40,000, assuming both Showcase participants do not overbid. Being called down to Contestant's Row is the winning lottery ticket by which one becomes a contestant on *The Price Is Right*, with good odds of winning valuable prizes.

In Contestant's Row, an item typically priced in the $500 to $2,500 range is presented for bid. Occasionally—for example, on special shows—items for bid can be priced considerably higher. The item may be displayed on stage (often durable goods), rolled across the stage (small vehicles), transported across the stage on *The Price Is Right* train, or descend from above (typically jewelry and small electronics). The four contestants bid on the actual retail price of the prize in sequential order, and the contestant bidding closest without going over wins that prize and comes on stage to play a pricing game. If a contestant guesses the exact price, he wins an additional $500 cash bonus, which occurred seventy-one times (3.3%) in seasons 47 and 48.

Figure 2.1 Contestant's Row Bidding—Episode 9,000, Season 48. *Getty Images*

Sequential order runs from left to right for the four contestants, with each empty seat being replaced by a new contestant from the audience. That new contestant bids first, with the bidding again going from left to right and wrapping back around. If all four contestants' bids are over the actual retail price, the bidding starts again in the same order.

Contestant's Row is the most strategically and mathematically complex part of the show. It is a sequential game with four contestants, each of whom has his own best estimate of the price of the item up for bids, and some semblance of a probability distribution around those estimates. Regular viewers of the show likely recognize the advantage of and strategy for bidding last. The last bidder can clip an earlier bidder by bidding just above a prior bid, thereby setting the high bid or choosing one of the two ranges on the number line between the first three bids. Or the last bidder can opt for the popular and dramatic $1 bid, wagering that the three earlier bidders have gone too high. By contrast, earlier bidders have to worry that a later bidder will take away some of the range from their preliminary claims or nearly all of the range in a clipping situation.

Viewers might observe that the highest bid disproportionately wins, particularly for more expensive items. Seasons 47 and 48 indicate a clear bias toward *underestimating* the value of the item up for bids. There is also evidence of later bidders being overly influenced by earlier bidders. For example, a low first bid may trigger a low second or third bid, as later bidders rebase their estimates. Separately, for strategic reasons, a later bidder may go high if earlier bidders have gone low.

CHAPTERS 2 THROUGH 5 ANALYZE CONTESTANT'S ROW IN FIVE STAGES

1. Results: What do the 2,136 instances of Contestant's Row bidding across 356 episodes reveal? How often does the first bidder win? And the last bidder? How often does the highest bid win and how does that vary with the price of the item up for bids?
2. Bad strategies—real-life examples: It is easier to learn based on wrong strategies before trying to expound on the correct strategies.
3. Qualitative observations: Based on the results, what conclusions can be drawn about good and bad bidding behavior? What does this say about our ability to strategize effectively and our psychological biases?
4. Game theory solution: I introduce game theory concepts for sequential games. I then create a simplified version of Contestant's Row bidding, a sequential game, to arrive at a discrete solution. Finally, I loosen the restrictions to arrive at qualitative strategies for each of Bidders 1–4.
5. The best strategies: Using qualitative observations and game theory analysis, I articulate strategies for astute bidding, from the last player back to the first. I use examples to highlight good strategic play for each of Bidders 1–4.

RESULTS

In seasons 47 and 48, there were 356 episodes broadcast on CBS All Access for a total of 2,136 instances of Contestant's Row bidding. With nine people in Contestant's Row on each show, there were a total of 3,209 contestants: 1,266 of the contestants were male and 1,898 were female, for a 40/60 split. The remaining 45 "contestants" were couples or families on five special episodes (Valentine's Day twice, pre-Christmas twice, and Manuela's Baby Shower episode).*

Bid order is the largest determinant of success, as seen in Figure 2.2. The fourth bidder, "Bidder 4," unsurprisingly performed very well, winning 40.6%

---

*There were five additional contestants (fourteen instead of nine) on a college football rivalry episode that aired November 27, 2019.

Bid Order	Wins	Share	<$750	$750-$999	$1000-$1499	$1500-$1999	>$2000
1	379	18%	20%	25%	17%	14%	15%
2	426	20%	20%	20%	20%	21%	19%
3	464	22%	22%	18%	22%	22%	24%
4	867	41%	37%	37%	42%	43%	41%
Observations	2,136		355	342	608	389	442
Over/Re-bid	12	0.6%					

Figure 2.2 Wins by Bid Order and Prize Value

of the time. "Bidder 3" won 21.7% of the time, modestly better than the second bidder, "Bidder 2" (19.9% win rate), who performed modestly better than the first bidder, "Bidder 1" (17.7% win rate). Bidder 2 and Bidder 3 each benefited some from mistakes made by earlier bidders.

Items up for bids had a broad price range. On the low end, about one in six items had a value less than $750. On the high end, nearly one in five had a value greater than $2,000, inclusive of special episodes (e.g., Big Money Week). Prize values averaged $1,458, with a median value of $1,299. The median is less affected by the higher priced outliers. The lowest prize value observed was $500, calling into question any bid below $500. The three highest priced items up for bids were (1) a $13,645 six-night trip to Monte Carlo on October 17, 2019; (2) a $6,749 Yamaha Upright Piano on January 24, 2020; and (3) a $5,700 Rolex watch on the December 31, 2019, New Year's special. There were another ten prizes in Contestant's Row priced between $4,000 and $5,100.

Bid position was most impactful when the item up for bids was priced in the average or somewhat above-average range (i.e., $1,000 to $2,000). Per Figure 2.2, when the prize had a value of between $1,000 and $1,999, Bidder 4 won 43% of the time, compared to 37% for prizes priced less than $1,000 and 41% for prizes priced more than $2,000. Interestingly, for prizes less than $1,000, Bidder 1 modestly outperformed Bidders 2 and 3. When the item up for bids was priced very high, Bidder 3 seemed to benefit at the expense of Bidder 1.

While win rates were mostly consistent across rounds, it is noteworthy that Bidder 4's win percentage was highest in round 2, at 45%, once he realized his leverage from bidding last, but before earlier bidders figured out how to counteract that leverage.

The other standout driver of bidding success, perhaps the most meaningful, is having the highest bid, which I designate as Bid Rank 1 (and the lowest as Bid Rank 4). An astonishing 53% of bidding rounds were won by the highest bid, whereas approximately 16% of wins went to each of the second highest, third highest, and lowest bids, as seen in Figure 2.3 and 2.4.

Bid Rank	Wins	Win Rate
1	1129	53%
2	329	15%
3	332	16%
4	346	16%
	2136	100%

Figure 2.3 Success of High and Low Bids

Bid Order	Wins with Highest Bid	Total Highest Bids	Win % with Highest Bid
1	129	303	43%
2	201	451	45%
3	250	524	48%
4	549	858	64%
Total	1129	2136	53%

Figure 2.4 Success of High Bids by Bidder Rank

Only Bidder 4 controls whether he is the highest bid, by virtue of bidding last. Still, Bidder 1, Bidder 2, and Bidder 3 each won more than 40% of the time when their bids ended up as the highest bid. Bidder 4 won an astonishing 64% of the time when he placed the highest bid.

Winning percentages were considerably lower for bidders submitting lower bids, and this held true for each of the four bidders. The second highest bid performed the worst, except for Bidder 4, who had his next best result as the second highest bidder. This reflects the highest bid often being $1 above the second highest bid, versus anything inherently problematic with the second highest bid.

Figure 2.5 shows two representative bidding events, with a prize value close to the median observed bid, and Bidder 4 winning with the highest bid that just clips the next highest bid. *Throughout this book, the bidders are listed in their order of bidding, not their left to right position.*

Finally, consider bidding slightly over a prior bidder ("clipping") and the celebrated $1 bid. I consider bids up to $5 more than a prior bid as clips. Quite a few players, to be considerate, clip an earlier bid by bidding $5 more. If a bid clips two prior bids, I classify it as having clipped the latter of the two bidders. For example, if Bidder 2 bids $900, Bidder 3 $901, and Bidder 4 $902, I count Bidder 4 as having clipped Bidder 3 and Bidder 3 as having clipped Bidder 2, but not Bidder 4 as having clipped Bidder 2. For $1 bids, I include bids of $100 or less, since the intent is the same—namely submitting the lowest bid and avoiding any risk of going over.

Show	Prize	Bidder 1	Bidder 2	Bidder 3	Bidder 4	Price
1/8/19 Round 5	24.2 MP Camera	666	875	999	**1000**	$1200
1/24/19 Round 5	2.3 GHz MacBook Pro	1100	1150	999	**1151**	$1299

Figure 2.5

Bidder 4's chances of winning are clearly maximized by clipping earlier bidders. Bidder 4 won 53.9% of the time when clipping earlier bidders (520 out of 965 rounds, as seen in Figure 2.6) versus 29.6% of the time when not clipping a prior bid (347 out of 1,171). Bidder 2 and Bidder 3 did not meaningfully improve their performance by clipping a prior bidder, given that Bidder 4 often did the same, drastically limiting their upside.

The $1 bid seems to generate more crowd applause than it does bidding success. Not surprisingly, $1 bids by Bidders 1–3 prevailed infrequently, but even Bidder 4 won only 26.1% of the time with a $1 bid, as seen in Figure 2.7. That was below his overall 40.6% win rate, and below the 44% win rate he would have achieved on those occasions had he instead clipped the highest bid by $1.

Whereas Bidder 4's superior options are to clip an earlier bidder or bid $1, he only clipped another bidder or bid $1 in 1,325 out of 2,136 bidding rounds (62% of the time). The other 38% of bids were inferior bids—namely, bids that categorically reduced the range of values he could win over. Notably, the frequency of inferior bids by Bidder 4 decreased as the show progressed, with only 31% of bids inferior in the sixth and final round of bidding. This is evidence of Bidder 4 learning as the show progresses.

The strategies for Bidders 1–3 are less straightforward. They lack the advantages of Bidder 4 and have to anticipate the likely actions of future bidders. Still, they can improve their odds if they can deliver bids that remain credible

Clipping prior bidder ($5 or less over prior bid)

Bid Order	Clipping Wins	Clips Bidder 1	Clips Bidder 2	Clips Bidder 3	Total Clips	Win %
2	18	79			79	23%
3	50	90	124		214	23%
4	520	201	273	491	965	54%
Total	588	370	397	491	1258	47%

Figure 2.6 Clipping Prior Bids

$1 Bids (includes bids $100 or less)

Bid Order	Wins	Total $1 Bids	Win %
1	1	7	14%
2	5	24	21%
3	9	54	17%
4	94	360	26%
Total	109	446	24%

Figure 2.7 $1 Bid Performance

but reduce the likelihood of being clipped by a later bidder. They can also bid higher, given that the highest bid prevailed more than half the time. Alas, they often do not prepare ahead of time because they do not know if they will be called down to Contestant's Row, in contrast to contestants on *Jeopardy*, who are chosen in advance.

The 1996 academic article "The Price Is Right, But Are the Bids? An Investigation of Rational Decision Theory," published in *American Economic Review*, discusses the topic of bidder learning in a more detail.[†] What was noteworthy in seasons 47 and 48 is that Bidder 4's win rate did not increase in the later rounds despite more optimized bidding, suggesting that Bidders 1–3 also became more effective in their bidding.

[†]Jonathan B. Berk, Eric Hughson, and Kirk Vandezande, "The Price Is Right, But Are the Bids? An Investigation of Rational Decision Theory," *American Economic Review* 86, no. 4 (Sept. 1996): 954–970.

3

Contestant's Row

Underbidding and Anchoring

The bidding results from seasons 47 and 48 presented in chapter 2 are illuminating. I wanted to allow the reader some space to form his or her own conclusions before offering my own hypotheses. Now let's dig into the reasons behind the results and discuss relevant examples.

BIDDER 4 IN POLE POSITION

The last bidder's chance of winning with strategic play must be at least 1/3, and potentially much higher. He always can bid $1 above Bidders 1, 2, or 3, or bid $1 if he feels strongly that the other bidders all overbid. By bidding $1 more than another bidder, Bidder 4 effectively reduces the contest from four to three bidders, since the clipped bidder cannot win short of getting the price right on the nose. Specifically, Bidder 4 should clip whichever bid he feels is closest to the actual retail price without going over. Even if he thinks all prior bids are over, he may still be best served by clipping a prior bid. If all bids are over, the bidding restarts in the same order, so Bidder 4 would again be able to bid $1 more than one of the first three bidders. That is not to say that Bidder 4 should never bid $1, just that there is no reason to do so unless he thinks his chance of winning is at least 1/3. Because Bidder 4's win probability with informed play is at least 1/3, Bidders 1–3 on a combined basis should win less than 2/3 of the time, which averages to less than 2/9 per bidder.

UNDERBIDDING BIAS

The highest bid winning 53% of the time is a startling statistic. Clearly this is not due to chance or solely due to the strategic interactions among the various bidders. It must be that bidders have a proclivity to underbid, which is *not* strategic in nature. Bidder 4 is positioned to take advantage of these downwardly biased earlier bids. Meanwhile, earlier bidders should clearly be bidding higher. Bidder 4 is not immune from underbidding either. Bidder 4 could have chosen to be the highest bidder in each of the 2,136 bidding contests by always bidding $1 more than the highest prior bid. He would have prevailed 50% of the time, significantly improving on his observed 40.6% win rate in seasons 47 and 48.

I observe seven principal reasons for underbidding, although there may be other causes.

- *Fear of going over.* The phrase "closest to the actual retail price without going over" instills a fear of going over. Contestants do not want to be disqualified from having their bid reviewed. However, this fear is illogical. Bidders 1–3 *should* be willing to risk bidding too high because their probability of winning is small, with respective win rates of 17.7%, 19.7%, and 21.7% in seasons 47 and 48. Each would have been well served to risk going over half the time if he could have won half the time when he did not go over (for a 25% win rate). The presence of four bidders enhances the risk/reward of bidding high. As a thought exercise, consider a hypothetical Contestant's Row with ten bidders instead of four. Here a bidder should have virtually no concern about overbidding, because his chances of winning are so low to start.
- *Bids based on sale prices*: Many people buy items on sale, not at retail. This biases bids downward, particularly for items often found on sale. In contrast, categories like electronics are often *not* found on sale, so contestants are less likely to underbid. Short vacations often trigger bids far below the retail price, as contestants are not accustomed to paying hotel rack rates. Some contestants seem to bid as if they are buying products rather than trying to win a game, thus amplifying the sale impulse.
- *High end brand/features*: Many familiar products like TVs still result in underbids because they include premium features that contestants are not accustomed to paying for. Other times the brand is more premium than what a contestants would purchase on their own dime.
- *Multiple-item mistakes*: Some prizes, like watches and luxury goods, include two to six items. The items are shown quickly, and the bidder must estimate a retail price quickly. Unless the bidder estimates the price per item separately, he may simply estimate a value for the whole

prize package rather than a value per item multiplied by the number of items. This approach usually leads to an underbid.

- *Unfamiliar items*: An unfamiliar item can trigger a broad range of bids, some meaningfully lower than the actual retail price, and some meaningfully higher. However, the low bids seem to outnumber the high bids as unfamiliarity typically leads to underestimation.
- *Bids less than $500*: There were no items for bid that were priced less than $500 in seasons 47 and 48. One should therefore always bid at least $500, except for $1 bids. Many contestants however, bid $400 to $500 on less expensive items.
- *Influenced bids*: Later bids are unduly anchored by earlier bids, causing underbidding to propagate to later bidders, regardless of the initial causes.

The implications of bids being biased low are many. For starters, the $1 bid loses its efficacy, because one is bidding less than bidders whose bids are inherently biased downward. This explains why Bidder 4 did worse with a $1 bid than he would have done as the highest bidder in those same bidding rounds.

It follows that Bidders 1–3 should have more success getting out of Contestant's Row when there are lower priced items up for bid, as was observed. Lower priced prizes rarely consist of multiple items and are often easier to approximate. The absolute dollar variation of such items tends to be lower, and bids tend to be closer to the actual retail price.

EXAMPLES OF UNDERBIDDING

There are hundreds of examples of underbidding that reflect the various biases discussed above. Because multiple biases are often at play in a given round of bidding, I highlight examples where a single bias stands out.

Fear of Going Over

A bid for an exercise bike had Bidders 1, 2, and 3 all bidding $1,000 or less, allowing Bidder 4 to bid $1,001, with an ample $298 cushion to the $1,299 actual retail price (see Figure 3.1).

One might have expected earlier bidders to anchor at or above the $1,000 level, as cardio equipment usually prices above $1,000. Relative to that $1,000 reference point, Bidder 1's $900 bid was a modest underbid, whereas Bidder 3's $600 bid was a significant underbid. Perhaps because of Bidder 1's low initial bid, Bidder 2 did not come in at a significant enough premium to avoid being clipped by Bidder 4.

Show	Prize	Bidder 1	Bidder 2	Bidder 3	Bidder 4	Price
4/2/19 Round 5	Exercise Bike	900	1000	600	**1001**	$1299

Figure 3.1

Bids Based on Sale Prices

Vacation bids are a good example of contestants underbidding based on sale prices versus retail prices. Vacations offered in Contestant's Row are more modest getaways than vacations offered in pricing games, typically two to four nights at a hotel or resort that is a short flight or limousine ride from Los Angeles. The lodging is usually the key component in the price of the vacation. The example in Figure 3.2 is for a four-night vacation to Lake Tahoe for two, including airfare and car rental.

Show	Prize	Bidder 1	Bidder 2	Bidder 3	Bidder 4	Price
1/25/19 Round 2	Lake Tahoe (4 nights + flight + car)	1450	1100	1300	**1451**	$2709

Figure 3.2

Here the average bid was roughly half the prize value, a discount that is fairly representative of bids for other vacation packages in Contestant's Row. In Tahoe, most hotels are connected to a casino, and hotel rates tend to be lower as a means of drawing customers to the casino. But *The Price Is Right (TPIR)* asks contestants to bid the actual retail price, not the sales price.

High End Brand / Features

Electronics often include premium features on everyday products, features that contestants are not informed about and are unlikely to buy. The example in Figure 3.3 is a fifty-five-inch OLED 4K TV for bid in late 2018, when OLED was not the standard.

Only Bidder 2 was at all close to the retail price. The others missed the added value of the OLED display.

Show	Prize	Bidder 1	Bidder 2	Bidder 3	Bidder 4	Price
10/17/18 Round 1	55" OLED 4K TV	1100	**2200**	1500	1150	$2800

Figure 3.3

Multiple Item Mistakes

In about one in seven bidding rounds, the item up for bids consists of multiple items, like a collection of watches or luxury goods. Some contestants make a bid that seems substantial in total but not on a per-item basis. Consider the example in Figure 3.4, with six Cole Haan luxury goods items.

Show	Prize	Bidder 1	Bidder 2	Bidder 3	Bidder 4	Price
2/1/19 Round 5	Cole Haan: 2x Jackets, Bags, Shoes	1200	1300	2000	**2001**	$2129

Figure 3.4

Bidder 1 and Bidder 2 placed bids equating to $200 per item, which seemed too low. Bidder 3 adjusted upward to a price of $2,000, or $330 per item, taking advantage of the downward bias of the first two bidders. Unfortunately for Bidder 3, Bidder 4 decided to go $1 higher and emerged victorious.

Unfamiliar Items

Contestants tend to bid too conservatively on unfamiliar items. For example, what does a Bluetooth-enabled jukebox cost? I am not sure, nor are most contestants. Perhaps as a result, the *highest* of the first three bids came in 40% below the actual retail price (Figure 3.5).

Show	Prize	Bidder 1	Bidder 2	Bidder 3	Bidder 4	Price
2/8/19 Round 1	Bluetooth Enabled Jukebox	600	800	900	**901**	$1496

Figure 3.5

A similar dynamic was observed on a show two days prior—the item up for bids was a turntable package with ten albums. The highest of the first three bids, $888, was more than 40% below the actual retail price of $1,523, with Bidder 4 bidding $900 and prevailing. When bidding on unfamiliar items, contestants should remind themselves that the median item up for bids was priced at $1,300, at least during seasons 47 and 48. They would be wise to anchor their bids for an unknown item around $1,300 instead of $1,000 or less.

The tendency to bid low on unfamiliar items can amplify other biases. For example, a contestant already fearful of going over may find this fear elevated if the item up for bids is unfamiliar.

Bids Less Than $500

Items up for bids were never priced less than $500 during seasons 47 and 48. This is not an accident, but a purposeful decision by *TPIR* to auction items with meaningful value. Any bid less than $500 is ill advised, and bids slightly more than $500 may also be too low (Figure 3.6).

Show	Prize	Bidder 1	Bidder 2	Bidder 3	Bidder 4	Price
4/7/19 Round 5	Indoor Games*	425	**450**	390	375	$531

Figure 3.6 Ping-Pong Table, Basketball Net, Golf Set, Rebounder

In the example in Figure 3.6, the indoor games had a very low $531 retail price. But the first three bids were all less than $500, with Bidder 2's $450 high bid prevailing.

The highest bid prevailed most frequently except when the item for bid was valued at less than $750 (see Figure 3.7). For sub-$750 items, the lowest bid (Bid Rank 4) prevailed 38% of the time, whereas the highest bid (Bid Rank 1) prevailed 25% of the time. Once the prize values exceeded $750, the highest bid was more likely to win than the lowest bid. And for expensive prizes, the success of the highest bid was nothing short of exceptional. The highest bid won 66% of bidding rounds for prizes priced between $1,500 and $2,000, and 81% of bidding rounds for prizes priced more than $2,000.

			Prize Value		
Bid Rank	<$750	$750-$999	$1000-$1499	$1500-$1999	>$2000
1	25%	32%	52%	66%	81%
2	11%	18%	19%	17%	11%
3	26%	23%	16%	11%	5%
4	38%	27%	13%	6%	3%
Observations	355	342	608	389	442

Figure 3.7 Win Frequency by Bid Rank, Prize Price

In the 1996 academic article, "The Price Is Right, but Are the Bids?" contestant underbidding was less pronounced. Although the sample size of 372 bidding rounds was small, contestants overbid 33.2% of the time then versus 24.2% during seasons 47 and 48. Surprisingly, 12.9% of bidding rounds in the 1990s had all four players going over versus 0.6% as aired in seasons 47 and 48. I believe about 5% of rounds had all four players going over, with most joint overbids not aired to create more time for commercials. Five percent of rounds rebid would translate to overbids occurring about 27.5% percent of the time. In the mid-1990s, Bidder 1 underbid far less, and other bidders likely followed his more aggressive lead. So why the difference today? Perhaps sale prices are more prevalent today than twenty-five years ago, whereas unfamiliar items up for bids show up more frequently. These factors likely cause Bidder 1 to bid cautiously, which in turn anchors other contestants.[*] Harvard professor Jonathan Hartley hypothesized about additional causes in his paper "Inattention and Prices over Time: Experimental Evidence from 'The Price Is Right.'"[†]

Undoing Underbidding

Underbidding on *The Price Is Right* has numerous implications for consumer behavior. If contestants on *TPIR* are inclined to underbid when it is not in their best interest to do so, imagine how much stronger these forces are for consumers in the real world. The causes of underbidding provide a blueprint for what sellers of goods and services need to *undo* in order to command near-retail prices. For starters, don't let buyers get too accustomed to sales prices. Perhaps discount for special events, but do not discount regularly. If the product is a high-end brand or has high-end features, make sure the consumer understands the brand value or the added value of the features. Many top-tier international brands cannot command a premium in the US market because they lack name recognition in the United States. Consider, if possible, having high-end features offered a la carte versus built into the product. Educate, expose, and explain, because an unfamiliar item will be discounted by consumers. When companies are being sold, bidders often increase their bids only after they are provided with additional information with which to value the businesses. Those in marketing for large corporations understand the aforementioned considerations and so much more, but for many small business owners, these strategies may not be as familiar.

OVERBIDDING AND INTRODUCTION TO ANCHORING

Because of underbidding biases, Bidders 1–3 should each be willing to risk going over more often than they do. Figure 3.8 shows how often each bidder overbids, excluding unaired overbids by all four bidders. Bidders 1 and 4 each went over 22% of the time, with Bidders 2 and 3 going over

---

[*]The mean prize surveyed in the 1996 article was nearly $1,000 versus $1,500 in seasons 47 and 48, but nearly twenty-five years of inflation would explain that difference.

[†]Jonathan S. Hartley, "Inattention and Prices over Time: Experimental Evidence from 'The Price Is Right'(1972–2019)," (October 11, 2019): http://dx.doi.org/10.2139/ssrn.3469008.

Bid Order	1	2	3	4	Total
Number of Overbids	477	559	575	468	2,079
Share of Bids Over	22.2%	26.0%	26.8%	21.8%	24.2%

Figure 3.8 Overbid Frequency by Bid Order (out of 2,136 Bidding Rounds)

nearly 27% of the time. Bidder 4's lower overbid rate may seem some-what counterintuitive, since Bidder 4 had the highest bid 40% of the time. But Bidder 4 is *not* often on the cusp of going over. Instead, Bidder 4 is taking advantage of underbids from Bidders 1–3 to clip the higher prior bid and winning 64% of the time when doing so (see figure 2.4).

Our simplified game theoretic model, discussed in chapter 4, suggests that bidders should collectively go over more than 40% of the time. Interestingly, if the item up for bids is less than $1,000, the aired "overbid rate" of 41% is in line with that model. Perhaps bidders do not expect items up for bids to be priced so low. For items that are more than $1,000, aired overbidding dropped off meaningfully, occurring 24% of the time for items between $1,000 and $1,499 and 11% of the time for items priced more than $1,500.

Notably, aired overbidding occurred at a suboptimal 23% level for the first five bidding rounds, but materially increased to 31% in the last bidding round. It is as if contestants realized they had one last chance to get out of Contestant's Row and finally started bidding more aggressively.

Analyzing overbids also sheds light on strategic interactions and how ear-lier bids influence later bids. Surprisingly, there were as many rounds with two or three overbids as there were with one overbid. *This should not occur if bids are independent (i.e., not influencing other bids).* The probabilities of a certain number of bidders overbidding independently is represented by the binomial distribu-tion and shown in Figure 3.9. That model predicts fewer instances of zero con-testants overbidding, more instances of one contestant overbidding, and fewer instances of three contestants overbidding. The probability of k overbids under the binomial formula is: $\binom{n}{k} p^k (1-p)^{n-k}$, where $n = 4$ is the number of bid-ders, k is the number of overbids, and $p = 24.2\%$ is the observed probability of overbidding in seasons 47 and 48. Here $\binom{n}{k} = \dfrac{n!}{k!(n-k)!}$.

Share of Rounds with Overbids

	0 Over	1 Over	2 Over	3 Over	4 Over	Total
Binomial	33.0%	42.2%	20.2%	4.3%	0.3%	100.0%
Observed	52.6%	15.3%	15.5%	16.1%	0.6%	100.0%

Figure 3.9 Frequency of Independent versus Observed Overbids

Low Bids Attract Low Bids

A higher observed frequency of zero bidders going over is strong evidence that low bids "attract" other low bids. More than half of bidding contests produced zero overbids, suggesting that underbidding propagates across bidders. However, if there is a broad range of possible prices for the item up for bids, a strategic player should want to carve out a different portion of the number than what has been claimed by prior bidders. Such an approach is even more beneficial in practice, given the high level of underbidding.

High Bids Attract High Bids

A high frequency of three bidders going over is strong evidence that high bids "attract" other high bids. About one in six bidding contests ended up with a triple overbid, versus 4% of independent binomial draws. Triple overbids are strong evidence of Bidder 1 (and to a lesser extent Bidder 2) influencing subsequent bidders, particularly against the backdrop of broad-based underbidding.

> **Anchoring Heuristics and Biases** could be the subject of an entire book. In the absence of information, individuals tend to orient toward the first number or numbers presented to them, however relevant or irrelevant those numbers may be. Psychologists Daniel Kahneman and Amos Tversky, in *Thinking, Fast and Slow*, shared an example of individuals who spun a wheel with random numbers between 0 and 100 (not the Big Wheel from *TPIR*) and were then asked how many countries are in Africa. Those who spun a high number gave high estimates, and those who spun a low number gave low estimates, despite the random number obviously having nothing to do with the answer. Anchoring is present everywhere in the world of stock investing. Investors often anchor to their purchase price as a gauge of when to sell a security, frequently and foolishly holding on to an underwater investment until it returns to their purchase price. Anchoring can affect a broad swath of market participants as well, causing a stock's trading multiples (e.g., price to earnings or enterprise value to earnings before interest, taxes, depreciation, and amortization) to remain at a discount or premium to peers based on past conditions. For example, if a company's business model has become less cyclical (positive) or outgrowth has slowed (negative), there is an opportunity for an investor to bet that the trading multiple will change and no longer be anchored to the past. Algorithmic trading by computers has likely amplified the duration of anchoring to past conditions. These trading multiples can be quite sticky even after many market participants realize the disconnect.

Key Tips

- Bidders 1–3 need to bid higher, and significantly so. They need to be willing to go over close to half of the time to meaningfully improve their odds of winning. Bidder 4 needs to bid $5 above one of the prior bids or bid $1 100% of the time.
- Bidder 4 would have been more victorious (50% vs. 40.6%) had he simply clipped the prior high bid every time. Bidder 4's always-bid-high strategy would not be valid if Bidders 1–3 did not meaningfully underbid. But if Bidders 1–3 continue to underbid, it is a very desirable strategy for Bidder 4.
- Situations in which contestants are more prone to underbid include unknown items, items typically on sale, items with premium features, and collections of items. Any bid below $500 is flawed, and bids slightly more than $500 are likely too conservative as well.

I address the question of how Bidders 1–3 ought to bid after developing a game theoretic model for Contestant's Row in chapter 4. "Higher is better" is a pretty good recipe to start with.

4

Bidding, the Game Theoretic Perspective

INTRODUCTION TO GAME THEORY

In game theory, sequential games are solved by looking forward and reasoning backward. Contestant's Row bidding is a sequential game, with a clear bidding sequence and strategies that are determined based on how subsequent bidders will react to prior bids. Other familiar sequential games include tic-tac-toe, *Connect Four*, and chess, although they involve numerous iterations, making them very complex. Bidding in Contestant's Row is also very different than simultaneous games like the famous "prisoner's dilemma" game, in which two prisoners guilty of jointly committing a crime and held separately in jail decide whether to confess their participation in said crime, with varying sentences.

Figure 4.1 shows a simple sequential game whereby Firm 1 is evaluating whether to enter an industry to take on a monopoly player. If Firm 1 does not enter, its payoff is zero because it has no investment and no gains. If Firm 1 does enter, the monopolist will fight back; for example, by lowering prices, which will lead to a negative outcome for both players, perhaps because the firms are losing money (e.g., airline fare wars). If the monopolist accommodates the new entrant, it still ends up with a payoff of "1," but not the "2" with no competitive entrant. Given the payoffs specified, Firm 1 should enter, because it knows that once it enters, the monopolist's best decision will be to accommodate it. However, in real life the monopolist might preemptively try to deter Firm 1 from entering; for example, by committing to fight before Firm 1 decides to enter. If that threat is credible, Firm 1 may decide not to enter.

We apply sequential game theory every day in our lives in games from cards to chess. The obvious sequential strategy in Contestant's Row is how

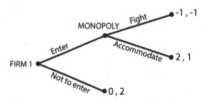

Figure 4.1 Illustrative Sequential Game

Bidder 4 should bid. He has four superior options—bid $1 more than one of the other three bids or bid $1 if he thinks his odds are maximized by betting that the other three bids are all too high. To make his decision, Bidder 4 considers his best guess for the actual retail price of the item up for bids, uncertainty around that price, and which of the four options maximizes his chance of winning. In effect, Bidder 4 has a rough probability distribution in his mind, which leads to estimated win probabilities for each of his four options. He then chooses the option with the highest probability. Bidder 4's probabilities add to nearly one, but not quite one, because one of Bidders 1–3 may have bid the exact price.

Since Bidder 4 can zero out a prior bidder's chance of winning by clipping that bidder, he can effectively reduce the game from four players to three players, ensuring himself of at least a one-third chance of victory. If none of the four options has a one-third chance of success, Bidder 4 should not bid $1, but instead clip one of the other three bids so as to maximize his chances of winning. For example, assume Bidder 1, Bidder 2, and Bidder 3 bid $600, $800, and $1,000, respectively. Now assume that Bidder 4 believes he has a 28% chance of winning with a $1 bid, and a 24% chance with a $601, $801, or $1,001 bid. Bidding either $601, $801, or $1,001 gives him a better than one-third chance of winning, *conditional on all four bidders not going over*. Why not bid $1? Because if all bidders go over, there is a second round of bidding, and Bidder 4 can assure himself of a one-third chance of winning or push bidding to a third round. And so on, ad infinitum.

Of course, each of the four bidders has a separate and distinct probability distribution, with a meaningful bias toward underbidding, as discussed in chapters 2 and 3. And estimates of the actual retail price by later bidders in the round may be dependent on the bids made by earlier bidders in the round, whether due to strategic considerations or the anchoring effect of earlier bids. A simple game like Contestant's Row bidding is actually far from simple.

A SIMPLIFIED GAME

To approach Contestant's Row bidding with rigor, I start with a simplified game with similarities to bidding in Contestant's Row.* In sequential order, each of four contestants bids on an item whose price is a random number between 1 and 100 (Figure 4.2). The winner is the one who bids closest without going over. To better illustrate the solution to the problem, I assume that one can bid out to two decimal values.

| 0 | *Bidder 1?* | *Bidder 2?* | *Bidder 3?* | *Bidder 4?* | 100 |

Figure 4.2

First, pretend there are only *two* players bidding. Bidder 2 will have a 100% chance of winning by way of clipping Bidder 1. Whatever Bidder 1 bids, Bidder 2 can bid an infinitesimally small amount higher. If Bidder 1 bids 40, Bidder 2 bids 40.01. Bidder 2 wins if the price is greater than 40.01, or both bids go over, and the game repeats. Bidder 2's strategy can be repeated ad infinitum until he wins.

Now consider a *three*-player game. Bidder 2 knows that Bidder 3 will bid an infinitesimally small amount more than either Bidder 2 or Bidder 1's bids or will bid near zero if the probability is sufficiently compelling versus the other possibilities. We do not know Bidder 1's ideal strategy a priori, but we can figure out how Bidder 2 should react and work backward to Bidder 1's strategy.

Assume Bidder 1 bids 50. Bidder 2 could then bid slightly higher, call it 50.01. But Bidder 3 would then bid 50.02, as shown in Figure 4.3, and Bidder 3 will have nearly ensured himself of victory or a redo if all bids are over.

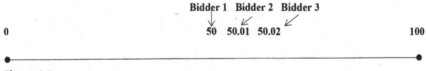

	Bidder 1	Bidder 2	Bidder 3		
0	50	50.01	50.02		100

Figure 4.3

If Bidder 2 bids 0.01 after Bidder 1 bids 50, Bidder 3 will prefer, just barely, bidding 50.01 versus 0.02. Bidding 50.01 gives Bidder 3 $49.99/99.99 = 49.99\%$ of the number line, whereas bidding 0.02 gives him $(50 - 0.02)/99.99 = 49.98\%$ of the number line (Figure 4.4). In the three-player game, Bidder 3 is guaranteed to claim at least half the number line between the lowest bid and 100, while having a 50% or better chance to win on the redo(s) if all three players go over.

---

*Although I derived the following results on my own, they parallel the results of Jonathan B. Berk, Eric Hughson, and Kirk Vandezande, "The Price Is Right, But Are the Bids? An Investigation of Rational Decision Theory," *The American Economic Review* 86, no. 4 (Sept. 1996): 954–70.

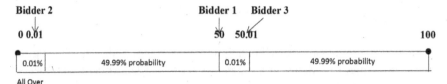

Figure 4.4

Can Bidder 2 do better after Bidder 1 bids 50? First let's define the **por**tion of the number line **o**ver **t**he **l**owest **b**id, (POTLB), which is all of the number line excluding the portion where all bidders are over. In the previous example, the POTLB is (0.01, 100); that is, all numbers greater than 0.01 inclusive of 100. Now back to Bidder 2. He knows that Bidder 3 can always claim at least half the POTLB by clipping bidder 1 or bidder 2 and that Bidder 3 has at least a 50% chance if all three bidders go over and are forced to bid again.

Bidder 2 can either go above or below 50. If he goes below, he is best served by bidding 0.01. How about if he goes above 50? A 50.01 bid would get clipped by Bidder 3, as would any bid that is not above 75. For example, if Bidder 2 bids 66.66, then Bidder 3 bids 66.67 and claims $(100 - 66.67)/50 = 2/3$ of the POTLB. Bidder 3 wins the first round with probability 1/3, and wins more than 50% of the time when a rebid is necessary, as he can clip Bidder 1 or Bidder 2 on the rebid. So with a 66.66 bid by Bidder 2, *Bidder 3's odds of victory are at least*: $1/3 + (1/2 \times 1/2) = 1/3 + 1/4 = 4/12 + 3/12 = 7/12$.

What if Bidder 2 bids 75.01? Then Bidder 3 can bid 50.01 and claim $(75.01 - 50.01)/50 = 1/2$ of the POTLB. Bidder 3 has a 25% chance of winning the initial bid, because half the time the random number is below 50 and all three contestants go over. He also has 50% chance of winning rebid rounds. *Here Bidders 3's odds of victory are at least*: $1/4 + (1/2 \times 1/2) = 1/4 + 1/4 = 1/2$.

Alternatively, Bidder 3 can bid 0.01 and have a $(50 - 0.01)/100 = 49.99\%$ chance of prevailing with a negligible chance of all three bidders going over. Here Bidder 3 ends up with the same near 50% probability, while Bidders 1 and 2 split the remaining 50%.

So Bidder 2's ideal response to a bid of 50 is to claim nearly 50% of the number line (i.e., probability) by bidding 0.01 and to let Bidder 3 claim 50% of the number line by clipping Bidder 1 with a 50.01 bid.

Notably, any bid by Bidder 2 above 50 but below 75 will encourage Bidder 3 to clip Bidder 2, zeroing out Bidder 2's chances. And any bid by Bidder 2 above 75 will still encourage Bidder 3 to bid 50.01. But Bidder 2 has a sub-25% chance of winning in the initial bid round with a bid above 75 versus a near 50% of chance of winning with a 0.01 bid. And Bidder 2 should have less than 50% odds of winning in the rebid that occurs half the time when all three bidders go over. So bidding 75.01 means a sub-50% chance of Bidder 2 winning and progressively lower odds for bids higher than 75.01.

Cleary Bidder 1 should not bid 50, as Bidder 2's best response is to bid 0.01, followed by Bidder 3 bidding 50.01. This is a recipe for failure for Bidder 1, shown again in Figure 4.5.

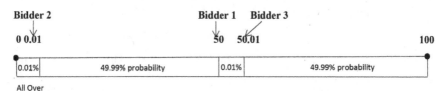

Figure 4.5

In practice, Bidder 2 might want to add an additional cushion and not be too greedy. That is, if Bidder 1 bids 50, then maybe Bidder 2 should bid something like 10 and not 0.01 so that it is really an obvious advantage for Bidder 3 to bid against Bidder 1 rather than Bidder 2.

It may be clear to readers by now that Bidder 2's strategy is to maximize his probability while guaranteeing Bidder 3 a 50% chance of victory, which has the effect of limiting Bidder 1's probability. So how does Bidder 1 bid so as to maximize his chances?

For starters, Bidder 1 knows that Bidder 3 will end up with at least 50% of the POTLB, because Bidder 3 can clip Bidder 1 or Bidder 2. Bidder 1 also knows that Bidder 2 can do no worse than Bidder 1 because Bidder 2 can react to Bidder 1's bid. It follows that Bidder 1 cannot create better than 25% odds of winning, assuming rational responses from Bidder 2 and Bidder 3. But can Bidder 1 bid guarantee a 25% chance of success? And if so, how?

We know 50 is a bad bid for Bidder 1, as it will likely be clipped by Bidder 3. If Bidder 1 goes slightly lower than 50, Bidder 2 will still bid 0.01, with Bidder 3 clipping Bidder 1's near-50 bid. If Bidder 1 goes far lower, say 20, then Bidder 2 will want to bid high, namely 60.01, with Bidder 3 bidding 20.01. Bidder 3 gets $(60.01 - 20.01)/80 = 50\%$ of the POTLB, while Bidder 2 keeps near half the POTLB for himself—$(100 - 60.01)/80 = 49.99\%$.

How about if Bidder 1 bids above 50? A 60 bid by Bidder 1 would cause Bidder 2 to bid 20.01 and Bidder 3 to bid 60.01, with Bidder 1 and Bidder 3 again splitting the POTLB (Figure 4.6).

But if Bidder 1 goes considerably higher, to 75.01, his situation improves. Bidder 2 could bid 50.01 and allow Bidder 3 to bid 75.02 or 50.02, but now

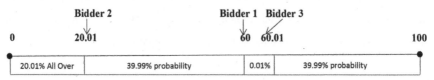

Figure 4.6

Bidder 3 is better off bidding 0.01 (Figure 4.7). Why? A 0.01 bid means a 50.01% chance of winning versus a 49.99% portion of the POTLB with a 50.02 bid or a 49.97% portion of the POTLB with a 75.02 bid. If Bidder 2 goes below 50, Bidder 3's best option is to clip Bidder 2 instead of bidding 0.01, as clipping now captures more than half of the POTLB.

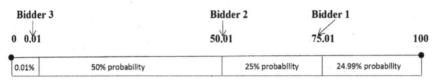

Figure 4.7

A 0.01 bid by Bidder 3 means that Bidder 1 captures 25% of the number line/probability, similar to Bidder 2. A bid above 75 by Bidder 1 would create a similar dynamic, with Bidder 2 bidding 50.01 and Bidder 3 bidding 0.01. Any bid above 75 by Bidder 1 just "donates" probability to Bidder 2.

In the three-player game, the best strategies for each bidder can be summarized as follows:

1. Bidder 1 bids 75.01, capturing 24.99%, or nearly 1/4 of the number line.
2. Bidder 2 follows with a 50.01 bid, capturing 25% of the number line.
3. Bidder 3 follows with a 0.01 bid and captures 50% of the number line.

Extrapolate to a four-player game

I mapped out the three-player game so as to avoid the complicated work of detailing the four-player game. In a four-player game, Bidder 4 is ensured of 33.3% probability of the POTLB, because he can clip Bidder 1, Bidder 2, or Bidder 3, and zero out one of these bidder's probabilities, effectively splitting the POTLB among three bidders.[†] Bidder 4 should consider bidding 0.01 (analogous to the $1 bid) only if he has a better than 33.33% chance of winning. Bidder 4's strategy is to maximize his probability with respect to the POTLB, because if

---

[†]To establish that Bidder 4 has at least a 33.3% chance of winning, consider the following: There are bids A, B, and C in lowest to highest order, with the price a random number between 0 and 1. Bidder 4's chances of winning by clipping A, B, and C, respectively, are $B - A$, $C - B$, and $1 - C$. Now Bidder 4's chances are minimized when clipping if he has an equal chance of winning from clipping each of Bidder A, B, and C, which means (1) $B = A + \frac{(1-A)}{3}$ and (2) $C = A + \frac{2(1-A)}{3}$, creating an equal incentive to clip each of the previous three bidders. Then bidding just above any of A, B, or C gives probability $\frac{(1-A)}{3}$ of winning the game immediately and probability A of replaying it. Assuming that Bidder 4's probability of eventually winning the game is P, this means that $P = \frac{(1-A)}{3} + AP$, or $P(1 - A) = \frac{(1-A)}{3}$, or $P = \frac{1}{3}$. Of course, if $A > 0.33$, then Bidder 4 may be better off bidding 0.01.

all bidders overbid, he has at least a 33.33% chance on the rebid, or the second rebid, ad infinitum.

Consider the following sets of strategies.

1. Bidder 1 goes high again, to capture 22.22%, or 2/9 of the number line with a 77.78 bid.
2. Bidder 2 follows with a 55.56 bid, also capturing 22.22% of the number line.
3. Bidder 3 follows with a 33.34 bid, capturing 22.22% of the number line.
4. Bidder 4 follows with a 0.01 bid and captures 33.33% of the number line.

This is an analogous solution to the three-player game and is shown in Figure 4.8.

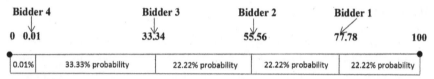

Figure 4.8

Can Bidders 1, 2, or 3 do better with alternative strategies? Collectively, they cannot do better than to limit Bidder 4's chance of winning to 33.33%. But can they do better versus one another?

Start with Bidder 3. If Bidder 1 bids 77.78 and Bidder 2 bids 55.56, then any bid by Bidder 3 above 33.34 reduces his chances of winning, as does bidding 55.57 or 77.79 (Figure 4.9). Bidder 3 slightly lowers his chances by clipping Bidder 1 or Bidder 2, as Bidder 4 can bid 0.01 and capture 55.55% of the number line.

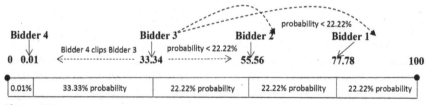

Figure 4.9

If Bidder 3 goes below 33.34, say 33, then Bidder 4 is better off clipping Bidder 3 with a 33.01 bid and capturing $(55.56 - 33.01/67) = 33.66\%$ of the POTLB. Bidder 4 knows that if all four bidders go over, he is assured of a 33.33% chance of winning on the rebid. So Bidder 3's best response to Bidders 1 and 2 is to bid 33.34.

Work backward to Bidder 2. Bidder 2 knows that Bidder 4 is assured of no less than a 33.33% or 1/3 chance of winning, by zeroing out one of the other bidders. Bidder 2 also knows that Bidder 3 can do no worse than a 22.22% or 2/9 chance of winning—at least 1/3 of the remaining 2/3 probability after subtracting Bidder 4's entitlement. If Bidder 1 bids 77.78, consider Bidder 2's options (Figure 4.10).

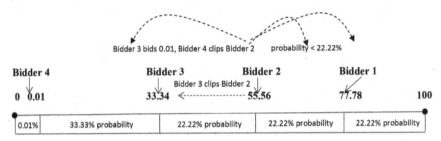

Figure 4.10

- If Bidder 2 bids 77.79, his probability of winning falls below 22.22%.
- If Bidder 2 bids below 55.56, say 53, he invites Bidder 3 to clip his bid with a 53.01 bid and to increase his chances to 24.78%. Then Bidder 4 bids 0.01 to capture $(53 - 0.01)/99.99 = 53\%$ of the POTLB.
- If Bidder 2 goes below 33.34, say 33.00, then Bidder 3 will bid 0.01, capturing $(33 - 0.01)/99.99 = 32.99\%$ of the POTLB. This leaves Bidder 4 to clip Bidder 2 with a 33.01 bid, capturing $(77.78 - 33.01)/99.99 = 44.77\%$ of the POTLB.

In summary, Bidder 2 going below 55.56 is self-defeating. A small deviation below 55.56 invites Bidder 3 to clip Bidder 2, while a large deviation downward invites Bidder 4 to clip Bidder 2 after Bidder 3 bids 0.01. So Bidder 2 can do no better than bidding 55.56.

Work backward to Bidder 1. Bidder 1 cannot do better than 22.22%, or $1/3 \times 2/3 = 2/9$ of the POTLB. Without going through all the scenarios that we did for Bidders 2 and Bidder 3, we know that Bidder 4 must get at least 33.33% or 1/3 of the POTLB. Bidder 3 needs to have at least as good odds as Bidder 1 and Bidder 2, and Bidder 2 at least as good odds as Bidder 1, given an ability to optimize bids in response to earlier bids. The 77.78/55.56/33.34/0.01 sequence by Bidders 1–4 satisfies these conditions.

Importantly, the sequence of the first three bids cannot proceed in a different order. Otherwise, as shown with Bidder 2 and depicted in Figure 4.11, either Bidder 3 or Bidder 4 will clip Bidder 2 and the other will go low (the same holds true for Bidder 1). Starting high allows Bidder 1 to *self-limit* his portion of the POTLB with a 77.78 bid, rather than letting other bidders clip him. Bidder 2 bidding 55.56 and Bidder 3 bidding 33.34 are also self-limiting actions. If Bidder 1 does not go high, some later bidder will find it in his best interests to clip Bidder 1.

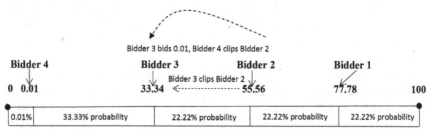

Figure 4.11

What is interesting is that the probabilities in the simplified game for Bidders 1–4 are not wildly out of line with the success rate achieved by actual contestants, with Bidder 4 doing better than 33.3% in practice and Bidders 1–3 each doing a bit worse than 22.2%. In real life Bidder 4 benefits from the errors made by Bidders 1–3, including underbidding. Nor is it surprising that Bidder 3 performed slightly better than Bidder 2, who performed slightly better than Bidder 1. Later bidders can take advantage of mistakes made by earlier bidders. I discuss these subjects as I generalize from the simplified game.

	Bidder 1	Bidder 2	Bidder 3	Bidder 4
Simplified Game	22.2%	22.2%	22.2%	33.3%
Seasons 47 and 48	17.7%	19.9%	21.7%	40.6%

The simplified game also calls attention to the observed underbidding bias. Although the simplified game solution almost never has all four bidders going over, by virtue of Bidder 4 bidding infinitesimally close to zero, it does imply that 42% of bids will go over. Note that random numbers from 0 to 100 will fall below Bidder 1's 77.78 bid 77.78% of the time, Bidder 2's 55.56 bid 55.56% of the time, and Bidder 3's 33.34 bid 33.34% of the time. On average, then, there are $0.7778 + 0.5556 + 0.3334 = 1.667$ overbids per bidding round, and $1.667/4 = 41.67\%$ of bids are overbids. Notably, in the simplified game Bidder 1 goes over more than Bidder 2, who goes over more than Bidder 3. This 41.7% overbid ratio compares to 24.2% in Seasons 47 and 48 (see figure 3.8), and perhaps 27.5% when estimating the number of instances where all four contestants overbid and it was not aired.

BIDDER 2 AND BIDDER 3 STRATEGY

Before generalizing from the simplified game, let us consider what happens if Bidder 1 fails to bid strategically (i.e., goes high) and instead bids his best spot estimate, 50, a pattern observed on the show. How should Bidders 2, 3, and 4 exploit this? First, note that there are two further cuts to the number line following Bidder 1's bid; namely those made by Bidder 2 and Bidder 3. Now, the best strategies for Bidder 2 and Bidder 3 are to position their bids so that Bidder 4 clips Bidder 1. In effect, Bidder 2 and Bidder 3 can split the POTLB three ways with Bidder 4.

How would this play out? If Bidder 1 bids 50, Bidder 2 can bid 25.01, and Bidder 3 can bid 75.01 (Figure 4.12). Bidder 4 clips Bidder 1 with a 50.01 bid to claim $(75.01 - 50.01)/74.99 = 33.34\%$ of the POTLB, while Bidder 2 and Bidder 3 claim an infinitesimally smaller $24.99/74.99 = 33.32\%$. So Bidder 4 clips Bidder 1 from above, and Bidders 2–4 share the POTLB 1/3, 1/3, 1/3. A 75.01 bid is indeed Bidder 3's best option. If Bidder 3 instead came in at 25.02, clipping Bidder 2, Bidder 4 would still bid 50.01. Bidder 3 would infinitesimally disadvantage himself and materially advantage Bidder 4. If Bidder 3 bid below 75, Bidder 4 would be able to claim more than 1/3 of POTLB by clipping Bidder 3.

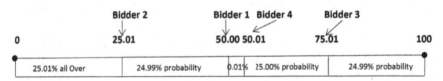

Figure 4.12

Interestingly, Bidder 2 could instead bid 75.01 and let Bidder 3 bid 25.01, with Bidder 4 again bidding 50.01 (Figure 4.13). The setup is essentially the same with Bidders 2 and 3 now taking opposite sides of Bidder 1.

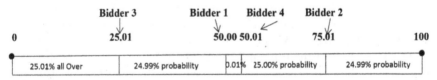

Figure 4.13

Nor would anything in the previous analysis substantially change if Bidder 1 underbid with a 40 bid. Bidder 2 and Bidder 3 could bid 10.01 and 70.01, or 70.01 and 10.01. Bidder 2 is still not disadvantaged versus Bidder 3 or Bidder 4, because he can move in such a way as to allow for equal cuts of the number line. If Bidder 1 were to bid somewhat above 50, say 60, then Bidder 2 would have to bid at least 73.34, self-limiting his chance of victory to 26.66%, so that Bidder 3 could bid 33.34, also for a 26.66% probability, and leave Bidder 4 with a 0.01 bid and a 33.33% probability (Figure 4.14).

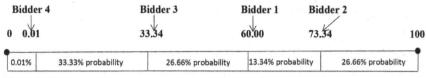

Figure 4.14

The key takeaway of this analysis is as follows: *unless Bidder 1 is bidding strategically high, Bidder 2 and Bidder 3 can materially improve their chances of winning by capturing a healthy portion of the number line while setting up Bidder 4 to clip Bidder 1.* This is perhaps the most critical insight into the complex matter of how Bidder 2 and Bidder 3 should bid. It leads to numerous corollary conclusions.

- If Bidder 1 bids in the "middle," Bidder 2 and Bidder 3 should take opposite sides of Bidder 1's bid, setting up Bidder 4 to clip Bidder 1.
- The higher Bidder 1 goes from the "middle" toward 77.78, the more Bidder 2 and Bidder 3 see their portion of the POTLB fall, from up to 33.33% to closer to 22.22% (i.e., four-player equilibrium), with one of Bidder 2 and Bidder 3 going above Bidder 1 and one going below.
- Underbidding by Bidder 1 does not change the ability of Bidder 2 and Bidder 3 to capture 1/3 of the POTLB. If Bidder 1 goes very low (below 33.34 in this example), then Bidder 2 and Bidder 3 should go high enough such that Bidder 4 will clip Bidder 1 and still capture at least 1/3 of the POTLB.

Now let us revisit the strategy of clipping, with reference to figure 2.6. Clipping worked well for Bidder 4, who clipped in 45.2% of bidding rounds and won 53.9% of the time when clipping. But clipping was far less effective for Bidders 2 and 3, who won only 23% of the time when clipping other bidders. The benefit Bidders 2 and 3 sought by clipping earlier bidders was often offset by their own bids being clipped: 26 of the 273 times Bidder 2 was clipped by Bidder 4 involved Bidder 2 clipping Bidder 1 and also being clipped by Bidder 4. And 82 of the 491 times Bidder 3 was clipped by Bidder 4 involved Bidder 3 clipping Bidders 1 or 2 and then being clipped by Bidder 4.

Even when Bidder 3 was not clipping an earlier bidder, his bids seemed to invite clipping by Bidder 4, to his detriment. Bidder 3 was clipped by Bidder 4 in 23% of rounds, but in only 1/6th of those rounds had Bidder 3 clipped an earlier bidder. In contrast, Bidder 1 and Bidder 2 were only clipped by Bidder 4 9.4% and 12.8% of the time, respectively.

GENERALIZING FROM THE SIMPLIFIED GAME

In the simplified game, I chose a uniform distribution between 0 and 100. The next step is to generalize to conditions more representative of Contestant's Row bidding.

The first generalization is to consider a generic probability distribution, be it a normal distribution (i.e., bell curve) or some other distribution. As long as the four players have *similar* probability distributions for the item up for bids, nothing really changes. Bidder 1 should bid to capture the priciest 2/9, or 22.22% of the probability distribution; Bidder 2 the next priciest 22.22%; Bidder 3 the third priciest 22.22%; and Bidder 4 the bottom 1/3, or 33.33%. If all players are rational and the probability distributions are the same, it need not matter what that distribution is.

The second generalization is to consider bidders with different probability distributions influencing one another. This is not straightforward. It will not be clear if the earlier player's bid reflects a best guess, a best guess with some cushion against going over, or a bid driven primarily by strategic considerations. On one extreme, the later bidder can assume the earlier player's bid was derived from the same probability distribution. An unexpectedly low or high bid would therefore be viewed as a strategic maneuver. On the other extreme, the later bidder can assume the earlier player's bid reflected a best guess tied to a different probability distribution.

I assume Bidder 1 provides his best guess with some cushion to reduce the likelihood of going over. Logically, this should be an appropriate characterization of Bidder 1 and for good reason. What else would Bidder 1 do without a knowledge of game theory but make a bid linked to his best estimate?

Consequently, a low bid from Bidder 1 may cause Bidder 2 to adjust his own "best guess with a cushion," which he then translates into a strategic bid. To illustrate such an adjustment, assume Bidder 1 bids $800 and Bidder 2's best guess with a cushion was set to be $1,400. Bidder 2 now makes an adjustment. Perhaps he adjusts to $1,200 based on one-third of the delta between Bidder 1's bid and his planned best guess. If Bidder 2 has more confidence in his own estimate, he may adjust less. If he has less confidence in his own estimate, he may adjust more.

Bidder 3 may also adjust his best guess and probability distribution in response to Bidder 2's bid. With that said, Bidder 2's bid is often driven more by strategic considerations (versus a best guess with a cushion). Bidder 3 is therefore more likely to adjust his bid in response to Bidder 1's bid, with confirmation or repudiation from Bidder 2.

Bidder 2 and Bidder 3, however, appear to adjust their bids too much. Statistics from seasons 47 and 48 support the notion that Bidders 2 and 3 follow Bidder

1 to an excessive degree when Bidder 1 bids materially too low or materially too high.

The standard deviation of bids 1–3 is a key metric to address this problem, defined $\sigma = \sqrt{\frac{1}{N}\Sigma(x_i - \mu)^2}$, where σ is the population standard deviation, N = the size of the population, x_i = each value from the population, and μ = the population mean. In effect, we are summing the squares of the differences from the mean bid, averaging that sum, and then taking the square root. Consider two examples.

1. Bidder 1 bids \$900 for a \$1,200 prize, Bidder 2 \$1,100, and Bidder 3 \$1,300. The standard deviation

$$\sigma = \sqrt{\frac{1}{3} \times ((900 - 1100)^2 + (1100 - 1100)^2 + (1300 - 1100)^2)} =$$

$\sqrt{\frac{1}{3} \times (40000 + 40000)} = \sqrt{\frac{80000}{3}} = 163.3$, with μ = \$1,100 based on the average of the \$900, \$1,100, and \$1,300 bids.

2. If Bidder 1 instead bids \$765 (15% lower), Bidder 2 bids \$935, and Bidder 3 bids \$1,105 (all 15% lower),

$$\sigma = \sqrt{\frac{1}{3} \times ((765 - 935)^2 + (935 - 935)^2 + (1105 - 935)^2)} =$$

$\sqrt{\frac{1}{3} \times (28900 + 28900)} = \sqrt{\frac{57800}{3}} = 138.8$, with μ = \$935 based on the average of the \$765, \$935, and \$1,100 bids.

The mean has been reduced by 15%, and the standard deviation has also been reduced by 15%. We can also calculate a relative standard deviation by dividing by the average of bids 1–3. We calculate a relative standard deviation for (1) of 163.3/1100 = 14.8% and for (2) of 138.8/935 = 14.8%. They are identical because the bids show a similar percentage dispersion.

Without undue influence, the standard deviation (absolute and relative) should increase if Bidder 1 bids low or high. Although Bidder 2 and Bidder 3 may be influenced by Bidder 1's bid, that influence should be modest. If the standard deviation of the three bids is nearly as low when Bidder 1 bids very low or very high, it is likely that Bidder 2 and Bidder 3 are being over-influenced by Bidder 1.

What we find is that for items for bid in the middle of the price range—\$750 to \$1,999, comprising 63% of bidding rounds—larger underbids by Bidder 1 not only lead to a lower absolute standard deviation, but a relative standard deviation that is essentially unchanged.[‡] In effect, the percentage dispersion is the same whether Bidder 1's bid is "typically below" the price of the

---

[‡]Note that the relative standard deviation is the more stringent test because the general bidding level is lower if Bidder 1 bids lower.

item up for bids, which we bucket as 15 to 30% below, or "atypically below" the price of the item up for bids, which we bucket as 30 to 45% below. *In other words, Bidder 2 and Bidder 3 are following Bidder 1 when Bidder 1 bids atypically low.* For very-low-priced items up for bids, it is unlikely that a bid will come in meaningfully below, whereas for very-high-priced items up for bids, a low bid may be more likely to be viewed as an outlier by Bidders 2 and 3.

For overbids, the standard deviation data are less conclusive as to whether Bidders 2 and 3 are being overly influenced by Bidder 1. Here we compare the percentage dispersion of the first three bids based on whether Bidder 1 is 0 to 15% above the price of the item up for bids or 15 to 30% above. Bidder 2 and Bidder 3 may be modestly influenced by a high bid by Bidder 1 (a rational response), rather than blindly following that high bid. Or perhaps there is both an influencing and an anchoring effect, and the anchoring effect is more impactful with low bids given a general underbidding bias.

Our analysis of overbidding does offer evidence of Bidder 2 and Bidder 3 following Bidder 1. Notably, three of four bidders went over *much* more frequently than the binomial distribution would suggest based on 24% of bids being overbids, as seen in Figure 4.15, or even the 6.0% expected if the true overbid ratio were 27.5%.[1] The frequency of triple overbids suggests later bidders are following high bids by Bidder 1. Zero overbids also were *much* more frequent than expected. This is consistent with a low bid by Bidder 1 influencing later bidders and making it very unlikely that one of them overbids.

Number Bidders Over	0	1	2	3	All 4	Total
Number of Rounds	1,129	329	332	346	12	2,148
Share of Rounds	52.6%	15.3%	15.5%	16.1%	0.6%	100.0%
Binomial Share of Rounds	33.0%	42.2%	20.2%	4.3%	0.3%	100.0%

Figure 4.15 Observed and Binomial Distribution of Overbids

Also consider standard deviation in the context of our simplified game. If Bidder 1 does not bid strategically and Bidders 2–4 do, a bid of 50 would trigger second and third bids of 25.01 and 75.01, for a standard deviation of 20, or 41%. If Bidder 1 bids 40, Bidder 2 and Bidder 3 should bid 10.01 and 70.01 (in either order), the standard deviation would increase to 24, or 61%. In other words, strategic play by Bidders 2 and 3 should cause both the percentage dispersion and the dollar dispersion to increase if Bidder 1 goes low. But in Seasons 47 and 48 we observe the opposite: the percentage dispersion stays nearly the same when Bidder 1 goes low, and the dollar standard deviation decreases. This is further evidence that Bidder 2 and Bidder 3 do not follow Bidder 1 for strategic reasons but rather are being over-influenced to their detriment.

[1] If 27.5% of bids are independent overbids, 0, 1, 2, 3, and 4 overbids would be expected to occur 27.6%, 41.9%, 23.9%, 6.0%, and 0.6% of the time, respectively.

Let's discuss examples of over-influenced bids before reviewing correct strategies for Bidder 2 and Bidder 3 in chapter 5.

On April 4, 2019, Bidder 1 came in very low for a pair of mountain bikes (Figure 4.16). Bidder 2 and Bidder 3 came in higher than Bidder 1, but not nearly high enough, and Bidder 4 clipped Bidder 2's high bid to comfortably win the $1,098 bikes with a $900 bid. It is hard to believe that neither Bidder 2 nor Bidder 3 were considering a bid on the order of $1,000 before Bidder 1's bid anchored them meaningfully lower.

Show	Prize	Bidder 1	Bidder 2	Bidder 3	Bidder 4	Price
4/4/19 Round 1	Pair of Mountain Bikes	650	899	800	**900**	$1098

Figure 4.16

High bids also encouraged other high bids. Consider the overbidding that occurred on January 15, 2019, for a 64 GB tablet (Figure 4.17).

Show	Prize	Bidder 1	Bidder 2	Bidder 3	Bidder 4	Price
1/15/19 Round 4	64 GB Tablet	800	900	799	1	$650

Figure 4.17

After a high $800 bid by Bidder 1, Bidder 2 went even higher. Bidder 3 then blundered by bidding $1 below Bidder 1 and, for purposes of this example, also bid too high. Bidder 4 made the easy $1 bid and won. Tablets are unlikely to cost more than $750 apiece, yet Bidder 2 and Bidder 3 followed Bidder 1 in overbidding.

Later bidders are less likely to be anchored lower or influenced higher when they have conviction about the price of the item up for bids or when Bidder 1's bid is so extreme that it is discounted entirely.

On July 4, 2019, a $700 bid by Bidder 1 for a 55-inch LED TV was doubled by Bidder 2 and almost tripled by Bidder 3 (Figure 4.18). Clearly, Bidders 2 and 3 were not influenced by Bidder 1's low bid, instead seeing the strategic opportunity to bid high and position themselves to win over a wide range of potential prize values. Unfortunately, they were too aggressive and may have overlooked the ever-declining trend of TV prices, allowing Bidder 4 to make an easy $701 bid for the win.

Show	Prize	Bidder 1	Bidder 2	Bidder 3	Bidder 4	Price
7/4/19 Round 5	55" 8500 Series 4K LED TV	700	1400	2000	**701**	$1300

Figure 4.18

On June 25, 2019, a bid of $900 for countertop appliances (coffee maker, rice cooker, and bread maker) by Bidder 1 led to more modest bids by Bidders 2–4 (Figure 4.19). Bidder 2 put forth a low bid that seemed to reflect conviction in a lower price. Interestingly, Bidder 1's approach was consistent with the simplified game solution for Bidder 1, aiming high so as to not be clipped, while preserving a decent portion of the relevant number line. This contest just did not fall Bidder 1's way.

Show	Prize	Bidder 1	Bidder 2	Bidder 3	Bidder 4	Price
6/25/19 Round 5	Countertop Appliances	900	450	600	**601**	$735

Figure 4.19

What is noteworthy in all four examples is that Bidder 4 won each time. For the pair of mountain bikes and 64 GB tablet, Bidder 2 and Bidder 3 followed Bidder 1 too low or too high, setting up an easy winning bid by Bidder 4. In the non–influenced scenarios, namely the LED TV and countertop appliances, Bidder 2 and Bidder 3 ignored Bidder 1 to such a degree so as to overcompensate in the other direction and create easy openings for Bidder 4 to clip and win. Being over–influenced is deleterious, because it leads to replicating bids that are too low or too high, which can then be exploited by Bidder 4. But being too confident in one's view can also backfire; for example, doing the opposite of Bidder 1 to such a degree so as to create a relatively easy win for Bidder 4 in the middle.

THE RATIONAL IRRATIONALITY OF BIDDER 1

A game theoretic approach is typically beyond Bidder 1's problem-solving skills. Nor does he recognize the benefit from bidding high and self-limiting himself. Instead, Bidder 1 typically reacts to the unresolvable complexity by bidding his best estimate of the price of the prize with some cushion. This is likely akin to choosing the average price over the zero to seventy-fifth percentile of the perceived price distribution. In fact, it is what Bidder 1 might do if we experimented with a simultaneous silent pricing game. In such a game, the Nash equilibrium would be a mixed strategy, with contestants randomly selecting a number in the zero to seventy-fifth percentile.§ But one player can simplify things by simply bidding the average across the zero to seventy-fifth percentile, provided the other players commit to the mixed strategy.

Consider an item whose price is drawn from the normal distribution with mean $1,000 and standard deviation $250. Let the contestant draw at random from the zero to seventy-fifth percentile of this distribution. The contestant will *on average* bid $900 and go over 40% of the time.** Of course the contestant could simply bid $900 straight up instead. This is analogous to a contestant bidding 37.5 for a prize uniformly priced between 0 and 100 versus choosing a number between 0 and 75 at random. By bidding $900 (the average of the zero to seventy-fifth percentile), the contestant creates a cushion against going over.

In summary, Bidder 1 often behaves as if bidding in a silent, simultaneous pricing game. But instead of choosing a number randomly from a probability distribution, which would be impractical, he chooses a number that is close to the probability weighted average of the zero to seventy-fifth percentile of the distribution. Call this the "simplified silent game solution" bid. It is his best estimate with a cushion, and it is a key consideration for Bidders 2 and 3, as I discuss in chapter 5.

Key Tip

The goal in Contestant's Row bidding is to maximize one's share of the portion of the number line over the lowest bid (POTLB). For earlier bidders, this is a function of capturing a decent, relevant portion of the number line and not being clipped. For Bidder 1, going high is the key self-limiting action to accomplish this goal. Bidders 2 and 3 should look for sizable unclaimed parts of the POTLB. Bidder 2 should consider the high option if Bidder 1 does not, whereas Bidder 3 needs to weigh the tradeoff of POTLB capture versus the risk of being clipped by Bidder 4.

---

§A Nash equilibrium is a proposed solution to a non-cooperative game involving two or more players in which each player is assumed to know the equilibrium strategies of the other players, and no player has anything to gain by changing only his own strategy. In a mixed strategy Nash equilibrium, no player has anything to gain in expectation by changing his strategy.

**We calculate the average of the inverse of the cumulative normal distribution function from $p = 0$ to $p = 0.75$. Because the zero to seventy-fifth percentile of the normal distribution is asymmetric, the average occurs at the fortieth percentile.

5

Contestant's Row Strategy—
Putting It All Together

Earlier chapters introduced game theoretic concepts, statistical observations, and principles on how and how not to be influenced by an earlier bidder's bid. Now we have a toolkit with which to outline general and situational strategies for each of the bidders.

First, let's review game theory strategies assuming similar probability distributions and non-influenced bids. This is representative of an item up for bids with a known and narrow price range; for example, an iPhone 10.

- *Bidder 4* should always bid $1 over one of the other bids or bid $1 so as to capture the largest probability of the portion of the number line over the lowest bid (POTLB). Data from seasons 47 and 48 suggest he could do even better by always tabling the highest bid.
- *Bidder 1* should bid very high in his probability distribution, ideally to a price that will stand as the high bid while capturing more than 20% of the probability in the POTLB. This entails going over more than half the time but winning about 20% of the time, usually with the high bid.
- If Bidder 1 goes high, *Bidder 2* should try and capture the next 20% or more of the POTLB by bidding below Bidder 1 but still reasonably high.
- *Bidder 3* should come in below Bidder 2 by a healthy amount but small enough such that Bidder 4 clips another bidder or bids $1.

Interestingly, there were only 44 bidding rounds out of 2,136 in which Bidders 1–4 were ordered from highest to lowest bids. Random ordering would suggest 1/24 of rounds should follow this sequence ($1/4 \times 1/3 \times 1/2$) versus the 1/49 observed. What I observed much more frequently, in a whopping 14.1% of the time (308 rounds), was the opposite pattern, with Bidder 1–4 ordered lowest to highest.

47

Bidding for designer sunglasses on December 10, 2018, was a rare example of near-perfect game theoretic play, apart from Bidder 3 not coming in low enough versus Bidder 2 (figure 5.1).

Show	Prize	Bidder 1	Bidder 2	Bidder 3	Bidder 4	Price
12/10/18 Round 6	4 Designer Sunglasses	2000	**1600**	1550	1	$1675

Figure 5.1

Second, consider the mental process by which Bidders 2 and 3 develop their best estimate of the retail price and translate that to a strategic bid, given that they may have different probability distributions and may be influenced by prior bidders. That process should work as follows.

1. Estimate their simplified silent game solution (SSGS) bid, in practice, their point estimate with a margin of safety. This is their "hypothetical first bid" and should be envisaged prior to Bidder 1's bid.
2. Evaluate how much to adjust their hypothetical first bid per (1) prior bids, (2) confidence in their hypothetical first bid, and (3) a view as to whether Bidder 1 was bidding his SSGS or bidding strategically. This leads to an "adjusted hypothetical first bid" (AHFB) and an adjusted probability distribution. This process is hardly exact and is particular to the item up for bids. Here Bidders 2 and 3 need to be careful not to be over-influenced by Bidder 1.
3. Translate their AHFB to a strategic bid; namely, going sufficiently higher or lower than Bidder 1 and/or Bidder 2, but with care not to go too much higher or lower.
4. Bidder 4 should go through the same process as Bidders 2 and 3, although the bidding strategy is simplified by deciding which of Bidders 1–3 to clip or whether to bid $1.

Time to discuss winning Contestant Row strategies!

"GO HIGH" STRATEGY

Game theory argues for Bidder 1 bidding high, with each player coming in sequentially lower, as seen in the designer sunglasses example. It also argues for Bidders 2 or 3 taking the high position if Bidder 1 or Bidders 1 and 2 fail to go high. The game theory solution is enhanced by a bias for contestants to materially underbid, such that the highest bid prevailed 53% of the time. Off-setting factors include the potential anchoring of later bids on earlier bids and the more average performance of high bids the 37% of the time the item up for bids was priced less than $1,000. As seen in figure 5.2, the high bid won 25% of the time for items priced less than $750, 32% of the time for items priced between $750 and $999, 52% of the time for items priced between $1,000 and $1,499, and 74% of the time for items priced more than $1,500.

Bid Rank	<$750	$750-$999	$1,000-$1,499	$1,500-$1,999	>$2,000	Total
1	89	111	315	257	357	1,129
2	40	60	115	65	49	329
3	92	79	96	43	22	332
4	134	92	82	24	14	346
Total	355	342	608	389	442	2,136

Figure 5.2 Wins by Bid Rank and Prize Value

We thus arrive at a general "go high" strategy. Each of Bidders 1–3 should try and go high in their probability distribution, provided that the item up for bids is likely priced more than $1,000 and provided that a prior bidder has not already bid high. Such bids should be high enough so as to limit the chances that one of the subsequent bidders clips their high bid. Also note that the $1,000 level will inflate over time from the 2019–2020 show data that is the basis for this book.

The April 3, 2020, contest for two paddleboards is a good example of Bidder 1 going high, and a good model for how Bidder 1 should regularly bid (figure 5.3). Although two paddleboards were unlikely to retail for more than $2,000, it was surely possible. Bidder 1 took a strategic bet by bidding high and won, with Bidders 2–4 believing their odds were maximized by bidding lower. Bidder 1, however, rarely bid this way, ending up as the high bid in only 1/7 of contests.

Show	Prize	Bidder 1	Bidder 2	Bidder 3	Bidder 4	Price
4/3/20 Round 5	Two 10 ft Inflatable Paddleboard	**2000**	1700	1150	1200	$2198

Figure 5.3

When Bidder 1 goes low, Bidder 2's best bet is to take the very high position that Bidder 1 should have taken, and same for Bidder 3 if both Bidder 1 and Bidder 2 fail to go high. The examples below in figure 5.4 show instances of Bidder 2 placing a winning high bid when Bidder 1 failed to go high, and Bidder 3 placing a high bid when Bidders 1 and 2 each failed to go high.

Show	Prize	Bidder 1	Bidder 2	Bidder 3	Bidder 4	Price
1/6/20 Round 3	3 Dior Sunglasses	950	**1200**	550	551	$1220

Show	Prize	Bidder 1	Bidder 2	Bidder 3	Bidder 4	Price
10/4/18 Round 3	Pair of iPads	800	1000	**1200**	1001	$1378

Figure 5.4

In the first example, Bidder 1 seemed to bid his SSGS (i.e., best estimate less a margin of safety). Bidder 2's bid was high enough to push Bidder 3's bid very low, with Bidder 4 choosing to clip Bidder 3. Bidder 2 prevailed, albeit with only a $20 cushion of the retail price. Examples like this are infrequent for Bidder 2, highlighting his room for improvement.

In the second example, Bidder 3 bid sufficiently high after Bidder 1's low bid and Bidder 2's moderate bid. Bidder 4 decided to clip Bidder 2 instead of Bidder 3, and Bidder 3 prevailed. For iPads, *sufficiently high* is not that much higher, given iPad pricing is known and narrow.

How high should Bidder 2 or Bidder 3 go in the absence of a prior high bid? Clearly it is a balancing act, executed well in the aforementioned examples. The bid needs to be high enough to keep the top spot but not egregiously high. The OLED TV example in figure 5.5 shows a more meaningfully high pivot after a low Bidder 1 bid. Even if the high bid was made with knowledge of OLED TVs and their higher prices (i.e., a higher AHFB), it was a smart bid strategically.

Show	Prize	Bidder 1	Bidder 2	Bidder 3	Bidder 4	Price
10/17/18 Round 1	55" OLED Smart TV	1100	**2200**	1500	1150	$2800

Figure 5.5

I believe that each of Bidders 1, 2, and 3 would improve his expected win rate by a few percentage points if he pursued a "go high" strategy and other bidders did not alter their strategies. The expected win rates of the other three bidders would correspondingly decline. Although we cannot go back and simulate bidding strategies for Bidders 1–3, we can do a "what if" exercise for Bidder 3 based on the bidding contests in which the item up for bids was priced more than $1,000 and both Bidder 1 and Bidder 2 *also* came in below the actual retail price.* For lower priced items, a strategy of bidding high could still have merit but should be evaluated on a case-by-case basis. *Notably, the "go high" strategy extends to Bidder 4, who should almost always bid $1 more than the prior high bid for items he believes to be worth over $1,000.*

In theory, Bidder 3 should see a larger improvement from the "go high" strategy than Bidder 2, who in turn should see a larger improvement than Bidder 1, as there are fewer subsequent bidders who might clip them. In practice, this may be less likely. Why? Because Bidder 4 was more inclined to bid above Bidder 3 than to bid above Bidders 1 and 2 and much more inclined to clip Bidder 3 when bidding above him. Figure 5.6 shows the twenty-four different rankings of bids by bidding order based on the rank of the first three bids and Bidder 4's response. Notably, Bidder 4 came in above Bidder 3 45.4% of the time when Bidder 3 had the standing high bid, but only 35.5% and 36.5% of the time above Bidder 2 or Bidder 1, respectively, when each of them had standing high bids. Moreover, when Bidder 4 came in above Bidder 3, he clipped Bidder 3 64.3% of the time, while clipping Bidder 2 only 56.5% of the time and Bidder 1 only 50.4% of the time when coming in above each of them.

Rank of Bids by Bidding Order, 1-3	4 above 3		4 above 2		4 above 1		4 low bid	Total	
3 > 2 > 1	302	(A)	109		65		168	644	(G)
3 > 1 > 2	134	(B)	61		61		60	316	(H)
2 > 3 > 1	108		110	(C)	49		83	350	(I)
2 > 1 > 3	76		138	(D)	50		85	349	(J)
1 > 3 > 2	85		31		97	(E)	54	267	(K)
1 > 2 > 3	55		34		77	(F)	44	210	(L)
Total	**760**		**483**		**399**		**494**	**2,136**	

(Bidder 4's Final Move)

Standing High Bids	Bidder 3		Bidder 2		Bidder 1	
Bids Topped	436	(A+B)	248	(C+D)	174	(E+F)
Total Bids	960	(G+H)	699	(I+J)	477	(K+L)
% of Bids Topped	45.4%		35.5%		36.5%	

Figure 5.6 Bid Orderings Based on Rank of First Three Bids and Bidder 4's Response

---

*In 48% of bidding contests (1,032 out of 2,136), the item up for bids was priced more than $1,000, and Bidder 1 and Bidder 2 also came in below the actual retail price. Bidder 3 bid higher in 559 of these 1,032 rounds (54%) and won 227 times, for a 40.6% conditional win rate. If Bidder 3 instead had come at a healthy premium to Bidders 1 and 2 in 75% of these rounds and prevailed 45% of the time, he would have won 16.2% of total bidding rounds (48% × 75% × 45%); 45% is a reasonable assumption given the 40.6% win rate, a likely better win rate if always bidding a healthy premium, and the 50% ceiling given Bidder 4's optionality. Now assume Bidder 4 also went higher when he should not have in 12% of total bidding rounds, either because the item was priced less than $1,000 or because Bidder 1 or Bidder 2 had overbid. And assume a 20% win rate in the remaining 40% of bidding rounds, Bidder 3's observed win rate for items up for bid priced less than $1,000. Bidder 3's win rate would have improved to 24.2% (16.2% + 20% × 40%), a 2.5-percentage point improvement over seasons 47 and 48.

Bidder 4's focus on Bidder 3 was so pronounced that Bidder 1 was not disadvantaged when making an accurate bid, which I define as between 20% below and 10% above the retail price. Each of Bidders 1–3 made an accurate bid about 30% of the time, and each won about 42% of the time when making accurate bids.[†] Driving this result was the fact that Bidder 3 was clipped 27.6% of the time with an accurate bid versus 23.0% of the time in total, while Bidder 1 was clipped only 20.4% of the time with an accurate bid versus 17.3% of the time in total.

The reasons for Bidder 4 more frequently coming in above Bidder 3 are threefold. First, Bidder 3 more often has the standing high bid among Bidders 1–3, as seen in figure 5.6 (boxed cells), which sets up Bidder 4 to come in above Bidder 3 when Bidder 3 has the high bid. Second, Bidder 4 likely assumes Bidder 3 has a more accurate bid by virtue of being able to incorporate information from Bidder 1 and Bidder 2's bids. This is evident by Bidder 4 also coming in higher than Bidder 3 more often when Bidder 3 had the second or third ranked bid. Third, Bidder 4 cannot see the prior bids and might forget such bids amid the excitement of the show and in formulating his own bid. (Contestants can ask Drew to repeat bids, but this is rare.) This last factor explains why Bidder 4 is less likely to clip Bidders 1 and 2 and why Bidder 3 does not stand to benefit more under the "go high" strategy than Bidders 1 and 2. *In contrast, Bidder 1 can table an accurate bid and not be disadvantaged. He is still better off going high, but targeting an accurate bid is not a bad strategy, particularly for lower priced items.*

BIDDER 4 STRATEGY

Bidder 4 has the easiest strategic setup but often failed to execute. He clipped Bidders 1, 2, or 3 by $5 or less only 45.2% of the time and bid $100 or less (usually $1) 16.9% of the time. The other 37.9% of the time Bidder 4's bid was neither $100 or less or within $5 of one of Bidders 1, 2, or 3. These were missed opportunities by Bidder 4 to capture a larger portion of the POTLB. In fact, if Bidder 4 had clipped 100% of the time while preserving his bid rank (i.e., highest, second highest, third highest, or lowest), he would have seen his winning percentage increase from 40.6% to 44.9%. *Clipping by Bidder 4 or bidding $1 is Bidder 4's superior strategy.* In fairness to Bidder 4, he may not always recall the prior bids, particularly amid the noise of the audience, thus limiting his ability to clip prior bidders. However, he can still ask Drew to repeat the prior bids, which few contestants chose to do.

[†]Bidders 1, 2, and 3 made accurate bids 30, 31, and 30% of the time, respectively, and won 41.7, 41.4, and 42.4% of the time when they did.

The Pilates reformer (exercise equipment for strengthening) on March 9, 2020, showcases Bidder 4's shortcomings (figure 5.7). Bidder 4 seemed comfortable that a $850 bid was low enough for the Pilates reformer, but he would have won by clipping Bidder 2 with a $701 bid. Perhaps the high $2,000 bid from Bidder 1 pushed Bidder 4's view of the price up a bit, causing him to not think twice about bidding $850.

Show	Prize	Bidder 1	Bidder 2	Bidder 3	Bidder 4	Price
3/9/20 Round 2	Pilates Reformer	2000	**700**	1050	850	$800

Figure 5.7

Given extensive underbidding, Bidder 4's win rate would have further improved to 50% if he had always clipped the prior high bid by $1. And Bidder 4 could have done even better than 50% had he followed this rule for items he believed to be priced more than $1,000 but considered clipping other bidders or bidding $1 for items priced less than $1,000, such as the Pilates reformer.

SITUATIONAL STRATEGIES AND EXAMPLES

There are many situations in which a high bid has already been tabled, and a bidder needs to respond accordingly, or the item's retail price is less than $1,000, and a "go high" strategy may or may not be in order. With that said, Bidder 1 finished with the highest bid only 14.2% of the time, and Bidder 2 with the highest bid 21.1% of the time, so do not underestimate the importance of the "go high" strategy.

Situational Play: Bidder 2

The simplified game solution suggests that Bidder 2 should table a moderately high bid after a high bid by Bidder 1. Consider such a course of action by Bidder 2. Bidder 2 won twenty-seven out of ninety-nine rounds when his bid ranked second and Bidder 1's bid ranked first, for a 27.3% win rate. In contrast, when Bidder 2 ended up as the third highest bid or lowest bid following a high bid from Bidder 1, he won only 12.5% of the time (11 out of 88 rounds), and 14.7% of the time (17 out of 116 rounds), respectively. Note that Bidder 2 does not have full control over his bid ranking second, so the relative win rates are directional rather than absolute.

The Apple Watch example from November 20, 2018, in figure 5.8 highlights sound situational play by Bidder 2. Bidder 2 undercut Bidder 1 but stayed high enough such that Bidders 3 and 4 went lower. Here Bidder 4 may have been influenced by Bidder 3's low $649 bid, choosing to stay low versus bidding $1,101.

Show	Prize	Bidder 1	Bidder 2	Bidder 3	Bidder 4	Price
11/20/18 Round 3	Pair Apple Watches, 1-year plan	1500	**1100**	649	900	$1388

Figure 5.8

Situational Play: Bidder 3

Bidder 3's job is easier than Bidder 2's but requires a delicate touch. What bid can he make that is sufficiently more than Bidder 1 and/or Bidder 2 or sufficiently less than the lower of the two, so as to maximize his probability on the POTLB while considering Bidder 4's likely response? Again, avoid the urge to clip. Whereas Bidder 3 won 23.1% of the rounds (50 out of 214) when he clipped—slightly better than his 21.7% overall win rate—this benefitted from Bidder 4 *not* clipping Bidder 3 35.7% of the time he came in higher than Bidder 3.

In response to a high bid by Bidder 2, Bidder 3 should go low if Bidder 1 has made a middle-of-the-road bid or in between Bidder 1 and 2 if he believes Bidder 1 has gone low. In both cases he should try to avoid being clipped by Bidder 4. Bidder 3's $900 winning bid for three wireless headphones on March 6, 2020, was the standing low bid (figure 5.9) but still high enough to encourage Bidder 4 to bid lower. (Bidder 4 should, of course, have bid $1 instead of $850.)

Show	Prize	Bidder 1	Bidder 2	Bidder 3	Bidder 4	Price
3/6/20 Round 4	3 Wireless Headphones	1100	1150	**900**	850	$1050

Figure 5.9

The landscaping package example from November 9, 2018, shows Bidder 3 skillfully and successfully bidding in between Bidder 1 and Bidder 2 without being clipped by Bidder 4 (figure 5.10).

Show	Prize	Bidder 1	Bidder 2	Bidder 3	Bidder 4	Price
11/9/18 Round 5	Landscaping Package	1250	900	**975**	1251	$1208

Figure 5.10

Still, bidding between a high and low bid was rarely successful for Bidder 3. When it was, it involved going between two bids that were neither too close together, so as not to limit Bidder 3's share of the POTLB, nor too far apart, lest Bidder 4 clip Bidder 3. Bidder 3 often erred, however, by bidding between Bidder 1 and Bidder 2 when their bids were not that far apart, in effect assuming an unwarranted degree of product knowledge and pricing precision by the first two bidders.

If Bidder 1 and Bidder 2 both are clustered in the middle, then Bidder 3 must decide whether to go low or high. Which action gives him a larger portion of the POTLB, but a portion not so large so as to be clipped by Bidder 4? This depends on whether the item is likely priced higher than $1,000 and the positioning of the other bidders. Generally speaking, Bidder 3 should bid *sufficiently but not excessively low* if Bidder 1 and Bidder 2's bids are high and *sufficiently but not excessively high* if Bidder 1 and Bidder 2's bids are low or middle of the road.

GENERAL SITUATIONAL STRATEGIES

Bidding decisions need to be made quickly by bidders who have different probability distributions and are incorporating information from prior bids. Still, it is useful to generalize the following scenarios, which combine situational strategies with the "go high" strategy.

a. If Bidder 2 and/or Bidder 3's adjusted hypothetical first bid (AHFB) is close to Bidder 1's bid, and the item is likely more than $1,000, then Bidder 2 should go sufficiently higher than Bidder 1 and Bidder 3 sufficiently lower, leaving Bidder 1 in the middle. If the item is likely less than $1,000, Bidder 2 should go sufficiently higher *or* lower than Bidder 1, with Bidder 3 taking the reverse position. In each case, place bids that make it more likely for Bidder 4 to clip Bidder 1.

b. If Bidder 2 and/or Bidder 3's AHFB is meaningfully higher than Bidder 1's bid, then they should each go higher, Bidder 2 much higher and Bidder 3 moderately higher. Again, bids should be calibrated to encourage Bidder 4 to clip Bidder 1.

c. If Bidder 2 or Bidder 3's AHFB is meaningfully less than Bidder 1's bid, Bidder 2 should bid modestly higher than his AHFB (but below Bidder 1) and Bidder 3 modestly lower than his AHFB, incentivizing Bidder 4 to bid $1.

Note that Bidder 2 and Bidder 3 do not necessarily have the same AHFB or probability distribution, so Bidder 2 may go meaningfully higher than Bidder 1, thinking Bidder 3 will go lower, only to find Bidder 3 going even higher.

As mentioned, for items likely priced less than $1,000, Bidder 2 may not be best served by going high even if Bidder 1 fails to go high. Consider the bidding for a dishwasher on February 2, 2020. Bidder 2 came in below a middle-of-the-road $799 bid by Bidder 1, which caused Bidder 3 and Bidder 4 to both clip Bidder 1 from above. Although a $700 bid by Bidder 2 might have further lessened Bidder 2's chances of being clipped, the $650 bid was directionally correct and prevailed (figure 5.11).

Show	Prize	Bidder 1	Bidder 2	Bidder 3	Bidder 4	Price
2/5/20 Round 4	Dishwasher	799	**650**	800	801	$750

Figure 5.11

Strategic bidding by Bidders 1–3 is no easy task, even if the examples here make it seem straightforward. My prescriptive strategies depend on the way that the other contestants play. Early bidders have to consider possible errors by later bidders when deciding what to bid, while keeping in mind that every deviation from the game theory prescription opens up opportunities for Bidder 4 to take an increasing share of the pie. Bidder 4, on the other hand, has four plausible bid options and can decide which option offers the best odds.

CORRECTING FOR UNDERBIDDING BIAS

On *TPIR*, there is generic underbidding and case-specific underbidding. Generic underbidding relates to the fear of going over and being accustomed to sale prices. Case-specific underbidding relates to specific properties of items for bid that trigger low bids, including high-end brands, multiple items in a bid package, and unfamiliar items. Generic underbidding requires standard adjustments, whereas case-specific underbidding requires focusing on the detail at hand and adjusting one's bid accordingly.

To determine an appropriate adjustment for underbidding, consider the simplified game, which suggests overbids should account for 42% of bids.[‡] I sought to answer the following question: *what percentage should contestants increase their bids in tandem to bring the level of overbids up to the level of the simplified game?* I made this adjustment for items that were priced less than $2,000, since very expensive items, whether jewelry or trips, lead to even more pronounced and idiosyncratic underbidding.

[‡]Although the simplified game assumes the price is a random number between 1 and 100, the optimal frequency of overbidding should be similar under most other probability distributions.

With a positive 10% adjustment for items less than $2,000, the frequency of overbidding would increase from an average 1.1 overbids per bidding sequence to 1.56 (i.e., 39% of contestants overbid). It follows that each player increasing his SSGS (or AHFB) by 10% is a good rule of thumb to offset the underbidding bias for sub-$2,000 items. However, given that most joint overbids did not air, I am inclined to suggest an 8% bump in lieu of a 10% bump. Of course, the win rate by contestants does not materially change if all players adjust their bids; the real benefit is for one contestant to correct for the underbidding bias when other contestants do not. Unfortunately, one cannot test just one bidder adjusting his bid upward by 10%, because his actions would likely cause later bidders to also adjust their bids.

A couple of additional pointers. First, items worth less than $1,000 generated a strategic 41% level of overbidding. Players should adjust for underbidding biases on a sliding scale, with negligible or no adjustments for lower priced prizes and higher adjustments for higher priced prizes. Second, Bidder 4 would have won 50% of bidding rounds had he consistently bid $1 more than the prevailing high bid.[§] Bidder 4's adjustment for underbidding is to clip the standing high bid except for lower priced items.

Key Tips

For items likely priced more than $1,000 (a level that will inflate over time), a "go high" strategy is in order. *Bidder 1* should go high. If *Bidder 1* fails to go high, *Bidder 2* should go high. *Bidder 3* should go high if both Bidder 1 and Bidder 2 fail to do so. If Bidder 1 goes high, *Bidder 2* should go moderately high. *Bidder 3* should bid *sufficiently but not excessively low* if Bidder 1 and Bidder 2's bids are high and in between Bidder 1 and Bidder 2 if Bidder 1 and Bidder 2 are high and low or low and high. *Bidder 4* should clip the prior high bid.

For items likely priced less than $1,000, *Bidder 1* should bid his best guess with a cushion, with *Bidder 2* and *Bidder 3* taking *sufficiently but not excessively* higher or lower positions based on their best estimates and where they see their odds maximized. *Bidder 4* should clip or bid $1.

Bidders 1–3 should increase their planned bids by 8% on average to counteract the underbidding bias, more for pricier items and negligibly for sub-$1,000 items. Case-specific causes of underbidding are better remedied with case-specific adjustments.

[§]Although 1,129 out of 2,136 rounds (53%) were won by the highest bid, 64 rounds (3%) featured an exact bid by Bidder 1, 2, or 3.

Broadening one's bid is a useful strategy for more than just Contestant's Row bidding. When one applies for college and business school, he is well-served to introduce other elements to his application, particularly if he has a standard background of good grades and good test scores from a good school or typical pre-MBA employment in the case of business school. A good standard background does not differentiate a candidate, especially since one's candidacy is evaluated across multiple features. A standard background is like a good point estimate bid in Contestant's Row. Chances are there will be a candidate with a better standard background who will diminish your prospects. But an applicant who can introduce compelling volunteer work, leadership experiences, entrepreneurial pursuits, or success in overcoming adversity often leapfrogs a generic but perfectly acceptable candidate. The same goes for inexperienced candidates in a crowded political race. A more experienced candidate will clip candidates who try to compete on experience. Better to differentiate your bid with a differentiated pitch.

6

Showcase Showdown

My young kids love the Big Wheel in the Showcase Showdown. "Big kids" love it too. Spinning the Big Wheel is colorful, competitive, and full of chance. Three players spin in sequential order, and each player tries to get as close as possible to $1.00 without going over in one or two spins. The contestant who has won the most combined between Contestant's Row bidding and his pricing game spins last, with the individual who has won the least spinning first, and the middle winner spinning second. The wheel must always go all the way around once for a spin to count, essentially making all spins independent. The first spinner's first spin starts on the $1.00 spot and all subsequent spins start from the spot of the last spin. The wheel contains twenty numbers in five-cent increments ranging from $0.05 to $1.00, arranged around the wheel as follows:

100	15	80	35	60	20	40	75	55	95	50	85	30	65	10	45	70	25	90	5

The contestant who does not exceed $1.00 and has the highest spin total across one or two spins goes to the Showcase. A tie goes to a one-spin spinoff, or a second spinoff if still tied after the first spinoff, with the wheel starting from the spot of the last spin. The Big Wheel is spun after the third and sixth pricing games, sending two lucky individuals to the Showcase.

If a contestant gets $1.00 in one spin or a combination of two spins, he wins $1,000 and gets a bonus spin, which starts at the $0.05 spot. If the bonus spin lands on the $0.05 or $0.15 spaces (the two green spaces), the contestant wins an additional $10,000. If the bonus spin lands on the $1.00 space, he wins an additional $25,000. In 712 Showcase Showdowns with 2,136 contestants, 182 contestants (8.5%) totaled $1.00 in one or two spins. In the bonus spin, 17 out of 182 players landed on $0.05 or $0.15 for a $10,000 bonus, and 8 landed on $1.00 for a $25,000 bonus.

Figure 6.1 The Big Wheel: Showcase Showdown, Season 38. *Getty Images Showcase Showdown Image*

Although the Big Wheel feels entirely like a game of chance, both the first and second players must choose whether to spin again or stop after their first spin. The contestant who spins last has the advantage of knowing what total he must beat or match.

Most players decide when to spin again and when to stop intuitively, and most make the correct probabilistic decision. It seems fairly obvious that Spinner 1 would not want to spin again with a $0.75 first spin, since his second spin would put him over 75% of the time. Similarly, a $0.25 first spin by Spinner 1 should lead him to spin again. He knows that Spinner 2 has not one but two chances to meet or beat the $0.25, as does Spinner 3.

Spinner 2's decision making is a function of how Spinner 1 performs. If Spinner 1 goes over, Spinner 2 must decide whether to stop after one spin or spin again. If Spinner 1 has not gone over, Spinner 2 must first match or eclipse Spinner 1's total. If Spinner 2's first spin exceeds Spinner 1's total, he must decide whether to spin again. Spinner 3 simply spins to beat or tie the higher total from Spinner 1 or Spinner 2. If both have both gone over, he spins once to try for $1.00 and the opportunity for a bonus spin.

I first discuss Showcase Showdown results and then introduce Monte Carlo simulation as a powerful tool to determine the correct strategies for Spinner 1 and Spinner 2. For readers versed in calculus, the Supporting Math appendix at the end of this book uses calculus to solve for the correct strategies, arriving at the same answer.

Showcase Showdown results are shown in Figure 6.2. Surprisingly, Spinner 2 prevailed as often as Spinner 3, notwithstanding the fact that Spinner 3 gets to "mindlessly" react to Spinner 1 and Spinner 2. Spinner 3 is like the house in blackjack, except without the requirement that he "hit again" at sixteen or less. In contrast, Spinner 1 and Spinner 2 occasionally err by spinning a second time when they should not and by not spinning a second time when they should.

Spinner	1	2	3	Total		Spin-Offs	$1,000	$10,000	$25,000
Wins	207	253	252	712		68	182	17	8
Win %	29.1%	35.5%	35.4%	100%		9.6%	8.5%	0.8%	0.4%

Figure 6.2 Showcase Showdown Results

Spinner 3 even has a slight advantage if his first spin results in a tie. If he ties at $0.45 or less, spinning again will produce a higher total without going over more than half the time. If he ties at $0.55 or more, he is better off opting for a spinoff. A $0.50 tie is a wash in terms of spinning again.

Why did Spinner 3 not outperform Spinner 2? Mainly chance, and also Spinner 2 benefitting from occasional miscues by Spinner 1. As seen in Figure 6.3, Spinner 2 totaled $1.00 in one spin or a combination of two spins on 71 occasions, versus 51 for Spinner 1 and 60 for Spinner 3. Moreover, 41 of the 71 $1.00 totals occurred on Spinner 2's second spin, meaning that he tallied $1.00 on 41 of his 518 second spins (7.9%), versus 32 out of 543 second spins by Spinner 3 (5.9%) and 20 out of 543 second spins by Spinner 1 (3.7%). Still, the results suggest that the advantage for Spinner 3 versus Spinner 2 is rather minimal. Otherwise, chance would have been insufficient to even out the win rates.

	Spinner 1			Spinner 2			Spinner 3			Win Value	% of Wins
Value	Ending Value	Wins (incl. ties)	Win Rate	Ending Value	Wins (incl. ties)	Win Rate	Ending Value	Wins (incl. ties)	Win Rate	Excl. 1/2 Both Over	Excl. 1/2 Both Over
$ 0.05	0	0		0	0		0	0		0	0%
$ 0.10	3	0	0%	0	0		1	0	0%	0	0%
$ 0.15	0	0		4	0	0%	4	0	0%	0	0%
$ 0.20	9	0	0%	2	0	0%	5	0	0%	0	0%
$ 0.25	4	0	0%	5	0	0%	6	0	0%	0	0%
$ 0.30	12	0	0%	8	0	0%	9	1	11%	1	0%
$ 0.35	5	0	0%	9	0	0%	9	1	11%	1	0%
$ 0.40	4	0	0%	15	1	7%	9	2	22%	3	0%
$ 0.45	12	0	0%	15	0	0%	16	0	0%	0	0%
$ 0.50	20	3	15%	20	1	5%	19	2	11%	6	1%
$ 0.55	20	1	5%	16	1	6%	22	4	18%	6	1%
$ 0.60	22	1	5%	27	2	7%	25	6	24%	9	1%
$ 0.65	46	5	11%	36	9	25%	23	8	35%	22	3%
$ 0.70	56	12	21%	34	11	32%	27	8	30%	31	4%
$ 0.75	69	23	33%	59	21	36%	29	12	41%	56	8%
$ 0.80	50	18	36%	41	16	39%	44	23	52%	57	8%
$ 0.85	67	26	39%	54	37	69%	41	26	63%	89	12%
$ 0.90	55	31	56%	53	38	72%	50	36	72%	105	15%
$ 0.95	63	42	67%	61	51	84%	56	46	82%	139	19%
$ 1.00	51	45	88%	71	65	92%	60	53	88%	163	23%
Over	144			182			233				
1/2 Both Over							24	24	100%	25	4%
	712	207	29%	712	253	36%	712	252	35%	713	100%
Spin Again	427			517			543				

> 50% Win Rate at Stopping Value

Figure 6.3 Spin, Ending Value, and Win Value Distribution (Number of Outcomes)

Later spinners spun again more frequently, as they often needed to try and beat a strong result from a prior spinner. As a result, they also went over more. Spinner 1 spun again 427 times (60%) and went over 144 times (33.7%). Spinner 2 spun again 518 times (73%) and went over 183 times (35.3%). Finally, Spinner 3 spun again 543 times (76%) and went over 233 times (42.9%).

When not needing to spin again to overtake Spinner 1, Spinner 2 was more conservative than Spinner 1 in choosing to spin again. Why? Because if Spinner 1 is out of the picture, Spinner 2 must only consider the possibility of Spinner 3 exceeding his total. The same dynamic explains why Spinner 1 had a lower win rate, even for high spin totals. Spinner 2 and Spinner 3 each had a chance to beat Spinner 1's total. To put that point in context, consider that Spinner 1 won 39% of the time at $0.85, whereas Spinner 2 won 39% of the time at $0.80 and Spinner 3 won 41% of the time at $0.75.

The rightmost columns in Figure 6.3 show the winning bid (or tie bid in a spinoff). It separates the twenty-five instances in which Spinner 1 and Spinner 2 both went over, providing an automatic win for Spinner 3. This is shown in the "1/2 Both Over" row. Excluding the winning values in those twenty-five rounds is more representative of the distribution of winning tallies, as Spinner 3 does not need to beat any prior result.

OBSERVED SPINNING STRATEGY

Most of the time it is intuitively obvious to Spinners 1 and 2 whether they should spin again. Only in the middle of the range for the first spin—namely, between $0.50 and $0.70—does uncertainty set in. This ambiguity is more pronounced for Spinner 1. He must consider his chance at having the highest value after Spinner 2 and Spinner 3 have both had a turn.*

Figure 6.4 shows the decisions to spin again (and not spin again) by Spinner 1 and Spinner 2. Spinner 2 has a decision to make only if Spinner 1 goes over $1.00 or if Spinner 2's first spin puts him ahead of Spinner 1's total. This occurred in 268 Showcase Showdowns, or 37.6% of the time. Of course, Spinner 2 will spin again if his first spin is below Spinner 1's tally.

Notably, Spinner 1 alternated between spinning again and not spinning again when his first spin was $0.60 or $0.65 and, in some instances, spun again at $0.70 and $0.75 or stayed at $0.50. Per Figure 6.4, he stayed two out of fifty-two times at $0.50 and ten out of thirty-three times at $0.60. Above $0.60 he mostly chose to *not spin* again, staying twenty-five out of forty times

---

*In game theory it would be said that Spinner 1 has to consider Spinner 2 and Spinner 3's reaction functions.

	Spinner 1 (# of Times)			Spinner 2 (# of Times)			
Value After 1st Spin	First Spin	Spin Again	Not Spin Again	First Spin	Spin Again	Not Spin Again	
$ 0.05	33	33	0	12	12	0	
$ 0.10	23	23	0	6	6	0	
$ 0.15	33	33	0	7	7	0	
$ 0.20	42	42	0	7	7	0	
$ 0.25	27	27	0	5	5	0	
$ 0.30	44	44	0	13	13	0	Shaded To
$ 0.35	39	39	0	12	12	0	Reflect
$ 0.40	26	26	0	7	7	0	Contestant's
$ 0.45	31	31	0	7	7	0	"Majority"
$ 0.50	52	50	2	5	3	2	Decision
$ 0.55	37	37	0	13	12	1	
$ 0.60	33	23	10	11	4	7	
$ 0.65	40	15	25	14	1	13	
$ 0.70	33	2	31	11	0	11	
$ 0.75	46	2	44	22	0	22	
$ 0.80	30	0	30	16	0	16	
$ 0.85	43	0	43	24	0	24	
$ 0.90	25	0	25	24	0	24	
$ 0.95	42	0	42	25	0	25	
$ 1.00	33	0	33	27	0	27	
Total	712	427	285	268	96	172	

Figure 6.4 Spin Again Decisions by Spinner 1 and Spinner 2

at $0.65, thirty-one out of thirty-three times at $0.70, and forty-four out of forty-six times at $0.75. Spinner 2 showed more consistency in deciding whether to spin again, mostly spinning again at $0.55 and below and mostly staying at $0.60 and above. The "zone of ambiguity" seems to be in the $0.60 to $0.75 range for Spinner 1 and the $0.50 to $0.65 range for Spinner 2.

SOLVING THE WHEEL

The most straightforward way to solve the wheel is to run a Monte Carlo simulation, which means running thousands of simulation trials using Microsoft Excel or other software. In Excel, generate a random number in $0.05-intervals between $0.05 and $1.00. Then create if/then statements based on the rules applicable to Spinner 1, Spinner 2, and Spinner 3, and repeat the formula across thousands of rows, each representing a simulation. Finally, sum up the results of who wins each simulated contest. Assume that all ties lead to spinoffs, with each Spinner winning 50% of two-way ties and 33% of three-way ties.

I first simulated Spinner 2 versus Spinner 3, which assumes that Spinner 1 has gone over or that Spinner 2 has surpassed Spinner 1 with his first spin. I ran three thousand trials that compared Spinner 2 stopping at each of $0.45/$0.50/$0.55/$0.60/$0.65 versus spinning again after those first spins. The relative win rates allowed me to identify the inflection point—the point at which Spinner 2 is better off not spinning again versus spinning again.

Figure 6.5 shows the simulation of Spinner 2 versus Spinner 3 (assuming Spinner 1 is no longer a factor) across three thousand trials. If Spinner 2 stops at $0.45, he only wins 19.9% of simulations, versus 29.5% if spinning again. If Spinner 2 stops at $0.65, he wins 39.8% of simulations, versus 23.3% when spinning again. The inflection point, where a second spin starts lowering the expected win rate, is at $0.55. At $0.55, Spinner 2's probability of winning is slightly higher by not spinning again. So Spinner 2 should spin again at $0.50 or less and stay at $0.55 or more, although spinning again at $0.55 is a close call.

1st Spin	No 2nd Spin Win Rate	2nd Spin Win Rate
45	19.9%	29.5%
50	24.4%	29.5%
55	29.6%	28.3%
60	35.5%	26.1%
65	39.8%	23.3%

Figure 6.5 Second Spinner Monte Carlo Simulation—Win Rate versus Spinner 3

The worst number for Spinner 2 to get in his first spin is $0.55, as spinning again and staying have roughly the same low odds of winning, just under 30%. In contrast, if Spinner 2's first spin is low, he can spin again with minimal risk of going over. And if Spinner 2's first spin is high, he can stay and have a decent chance of prevailing. Interestingly, the probability of winning after spinning again does not change much when the first spins range from $0.45 to $0.65, varying from only 29.5% to 23.3%. The probability of not spinning and winning is very sensitive to the first spin tally, increasing from 19.9% at $0.45 to 39.8% at $0.65.

Now let's simulate Spinner 1's strategy knowing how Spinner 2 will behave. Spinner 1's choice is less intuitive, as he must consider the probabilities of his total prevailing versus both Spinner 2 and Spinner 3. Here I ran three thousand simulated trials of Spinner 1 stopping at each of $0.50/$0.55/$0.60/$0.65/$0.70 versus spinning again in order to see what stopping point maximizes Spinner 1's odds of winning (Figure 6.6).

The simulation suggests that Spinner 1 is better off spinning again if his first spin is $0.65 or less. This is a higher value than the $0.50 stopping point I established for Spinner 2, as Spinner 1 needs to outperform two subsequent contestants. For Spinner 1, a good total becomes more important than the risk of going over.

1st Spin	No 2nd Spin Win Rate	2nd Spin Win Rate
50	5.9%	19.8%
55	9.2%	20.3%
60	11.6%	18.7%
65	16.0%	16.9%
70	20.9%	16.2%

Figure 6.6 First Spinner Monte Carlo Simulation—Win Rate versus Spinner 1 and Spinner 2

For Spinner 1, the worst first spin is $0.65. At $0.65 he should spin again, even though spinning again offers a paltry 16.9% expected win rate, which is below the win rates that can be expected when staying at $0.70 and more or when spinning again after a spin of $0.65 or less.

You may have noticed that the inflection points between spinning again and stopping for both Spinner 1 and Spinner 2 occur after getting the most challenging first spin value. At the inflection point the spinner's first spin value points to low odds of winning. Those odds remain low whether he chooses to spin again or stay after his first spin.

SIMULATING OPTIMAL STRATEGIES

Armed with the optimal stopping strategies for Spinners 1 and Spinners 2, I ran a ten-thousand-trial Monte Carlo simulation across the three spinners using optimal strategies, as shown in Figure 6.7.

Spinner	Spins at or Below	Outright Wins	Spinoff Wins After Ties	Total Wins	Total Losses	Going Over Losses	Win Rate
1	$0.65	2,874	281	3,155	6,845	2,237	31.6%
2	$0.50	2,963	308	3,271	6,730	2,900	32.7%
3	Reactive	3,165	410	3,575	6,426	3,134	35.7%
		9,002	998	10,000	20,000	8,271	100.0%

* Spinner 2 only spins at or below $0.50 if he need not beat a higher tally from Spinner 1

Figure 6.7 Simulated Win Rates with Optimal Stopping Strategies

In this simulation, Spinner 1, with a win rate of 31.6%, and Spinner 2, with a win rate of 32.7%, fall shy of winning one-third of the contests. Spinner 3, with a 35.7% win rate, wins more than one-third of the contests. I found it surprising that the simulated probabilities did not favor Spinner 3 more relative to Spinners 1 and 2. I expected a larger benefit from spinning third and reacting to the prior spinners' outcomes. Simulated ties occurred in about 10% of the contests, a similar rate to what was observed. Earlier spinners were almost as likely to lose by going over as they were to lose by having their totals eclipsed.

CONTESTANT PERFORMANCE VERSUS MODEL

During seasons 47 and 48, Spinner 2 performed 2.7 percentage points better than my Monte Carlo simulation of the optimal strategy. Spinner 1 performed 2.7 points worse than simulated, and Spinner 3 exactly in line. This is mostly due to chance, as spinner mistakes had a small impact. The probability distribution of win rates under optimal strategies does not converge fast, which is why the 712 real-life observations resulted in win rates moderately different than our ten-thousand-trial simulation. Even refreshing the ten-thousand-trial simulation led to meaningful variation in win rates. Across ten thousand trial runs, Spinner 1's win rate varied from 30.7% to 32.4%, Spinner 2's win rate varied from 31.6% to 33.2%, and Spinner 3's win rate varied from 35.1% to 37.1%. This is further evidence of slow convergence.

There was only slight divergence between the optimal strategy and contestant behavior on the Big Wheel. Spinner 1 and Spinner 2 are pretty rational in their decision making. Spinner 2 deviated from this strategy of staying at $0.55 and more in only 19 out of 268 contests (7%).[†] Spinner 2's mistakes included (1) staying twice at $0.50, (2) spinning twelve times at $0.55, (3) spinning four times at $0.60, and (4) spinning once at $0.65. Little was lost by these errors. They reduced Spinner 2's expected win rate 4.2 percentage points per errant contest, but only 0.3 points across the 268 contests, and only 0.1 points across all 712 contests. Relative to the observed 35.5% win rate and the 32.6% modeled win rate, 0.1 points is small.

A similar set of minor errors were made by Spinner 1, who erred in 41 out of 712 contests (6%). The average mistake reduced his chances of success by 3 percentage points on the contest in question but only 0.2 points averaged over all 712 contests—in other words, minimally versus the 29.1% observed win rate and the modeled 31.6% win rate. Most of Spinner 1's mistakes (35 out of 41) involved staying at $0.60 or $0.65 when he should have spun again. He also stayed twice at $0.50, spun twice at $0.70, and spun twice at $0.75 when he should have done the opposite.

Interestingly, 90% of Spinner 2's errors (17 out of 19) involved spinning again when he should have stayed (i.e., at $0.55 or more), whereas 90% of Spinner 1's errors (37 out of 41) involved staying when he should have spun again (i.e., at $0.65 or less). This suggests a slight bias to overspin on the part of Spinner 2 and a slight bias to underspin on the part of Spinner 1.

Given minimal contestant error, Spinner 1's low win rate mainly reflected chance. Spinner 1 spun $1.00 51 times versus an expected 57 times and won only 22 out of 50 spinoffs. He also spun between $0.50 and $0.70

[†]The 268 contests in which Spinner 1 went over or Spinner 2's first spin eclipsed Spinner 1's total.

more frequently than expected on his first spin, these spin values offering the lowest odds of victory. Conversely, Spinner 2 totaled $1.00 71 times in one or two spins versus an expected 61.5 times, aiding his chances versus Spinner 1 and Spinner 3.

Contestants are quite rational on the Big Wheel, suggesting a good intuitive grasp of probabilities in straightforward games like the Showcase Showdown. I attribute this to two factors.

1. *Simple math approximations*: Spinners have a good feel for when spinning again will put them at high risk of going over. And though they act conservatively because of this, they do account for the necessity of going for a high enough spin total to hold off subsequent spinners. For example, if Spinner 2 spins $0.60, he knows there is a 60% chance of going over on a second spin. He also knows there is a good chance of Spinner 3 beating his tally if he does spin again. Spinner 1's calculation is more complex, and he is more likely to err conservatively, given the need to hold off both Spinner 2 and Spinner 3.
2. *Experience*: Contestants who regularly watch the show may recall prior behavior on the wheel. Moreover, after players spin, Drew Carey sometimes calls out risk takers (i.e., those who spun a second time when the correct course of action was not spinning) or overly cautious contestants (i.e., those who did not spin a second time when the correct course of action was spinning).

So do not stress about the Showcase Showdown if you find yourself spinning the Big Wheel. First off, your instincts are mostly on target. And second, you can simply commit to memory the strategy of spinning again at $0.65 and less if Spinner 1 or $0.50 and less if Spinner 2.

Key Tip

Spinner 1 should spin again at $0.65 and less. Spinner 2, having surpassed Spinner 1, should spin again at $0.50 and less. If tied after your first spin, spin again at $0.50 or less.

7

The Mystery of Second Showcase Underperformance

The Showcase is the grand finale of the show. Each showcase is filled with amazing prizes usually totaling $20,000 to $40,000 but occasionally more. At least one of the showcases is presented based on some sort of theme that adds entertainment value, such as "the ride of a lifetime" or "things that will bring out the explorer in you."

The two winners of the Showcase Showdown face off in the Showcase. The contestant who has the higher dollar value of winnings leading up to the Showcase—the "winner"—sits on the left (the right as seen by the audience) and has the option of bidding on the first showcase or passing it and bidding on the second showcase. The contestant with the lower dollar value of winnings—the "runner-up"—sits on the right (the left as seen by the audience) and bids on the showcase that the "winner" chooses not to bid on. Whoever bids closest to his showcase value without going over wins his showcase.

A winning contestant who is less than $250 away from the value of his own showcase without going over wins both showcases. This happened twice in seasons 47 and 48, although it might have occurred more if not for four contestants going over by less than $100. If both contestants overbid, there is no showcase winner. This occurred about 6% of the time.

I approached the Showcase with some clear hypotheses. I expected there to be a higher win rate on the second showcase. The bidder on the second showcase can react to the bid on the first showcase, bidding closer to his estimate of the showcase value (i.e., "aggressively") or with a larger buffer to his estimate (i.e., "conservatively") as circumstances warrant. I also expected better performance on the lower priced showcase. The lower priced showcase should have a lower estimation error and may have two or three moderately priced

Figure 7.1 Veteran's Day Showcase, Season 42. Getty Images Showcase Image

prizes whose estimation errors offset rather than one car or boat comprising the vast majority of the showcase value.

What I found in the Showcase was not what I expected in terms of the win rates across first versus second showcases and lower priced versus higher priced showcases. Consider the following observations, which I then discuss in more detail.

- Showcases on *The Price Is Right* (*TPIR*) had an average value of $30,500 and a median value of $29,600, the higher average reflecting a few high outliers. Most high showcase outliers occurred on special episodes like those during Big Money Week. No showcase was ever less than $20,000.
- Showcase values were for the most part distributed on a Bell curve, albeit with high outliers and a larger cluster in the mid-high $20,000 range, and with no sub-$20,000 showcases. The showrunners generally avoided presenting two showcases that were very close in price.
- Surprisingly, the first showcase was disproportionately won, even if it was higher priced.
- The "winner" bid on the second showcase 66% of the time. He won less frequently because the second showcase was won less frequently. On average, the "winner" selected for higher priced showcases, but the higher showcase values did not compensate for the lower win rates.

- To understand the inferior performance on the second showcase, I separately considered lower priced and higher priced showcases. When the lower priced showcase was presented second, contestants were too aggressive, overbidding 45% of the time. The cause seemed to be an anxiety tied to bidding second, resulting in unnecessarily aggressive bids.
- The lower priced showcase won over half the time when presented first but lost more than half the time when presented second, limiting the advantage in win rate for lower priced showcases.

Figure 7.2 details showcase results on an overall basis (i.e., blending the two showcases on each show together), the first and second showcases separately, and the showcases bid on by the "winner" versus those bid on by the "runner-up." I removed the twenty-four Showcases featured in special

All Showcase Observations Less 24 Special Shows

	All	Wins	1st Showcase	2nd Showcase	Winner Position	Runner-Up Position
Average	$29,842	$29,787	$29,515	$30,168	$30,343	$ 29,340
Median	29,361	29,365	29,342	29,366	29,756	28,677
Standard Deviation	5,266	4,714	4,864	5,628	5,387	5,102
Max	50,488	46,328	45,923	50,488	50,488	46,650
Top Quartile (>)	32,793	32,748	32,489	33,342	33,479	32,352
Bottom Quartile (<)	25,885	26,573	25,771	25,885	26,445	25,587
Min	20,264	20,350	20,318	20,264	20,264	20,318
Average Underbid	3,299	3,888	3,854	2,743	3,293	3,404
Median Underbid	3,044	3,252	3,400	2,422	3,271	2,794
Stdev Underbid	4,699	3,134	4,702	4,622	5,140	4,986
Win	312	312	175	137	142	170
Win %			52.7%	41.3%	42.8%	51.2%
Loss: Other Player	312	N/A	137	175	170	142
Loss %			41.3%	52.7%	51.2%	42.8%
Loss: Double Over	40	N/A	20	20	20	20
% Double Over	6.0%		6.0%	6.0%	6.0%	6.0%
Avg Value of Win	$29,787	$29,787	$29,436	$30,234	$30,718	$ 29,009
Average Earnings	27,992		15,516	12,476	12,253	13,852
Over	160	N/A	62	98	83	77
% Over	24.1%		18.7%	29.5%	25.0%	23.2%
% Losses Over	45.5%		39.5%	50.3%	43.7%	47.5%

Figure 7.2 Showcase Statistical Summary

shows. Those Showcases were usually more expensive, averaging $39,800, or $10,000 more than the other 332 Showcases.* Although contestants would have welcomed being in the Showcase on these special episodes, such showcases are not representative and skew the results. *All showcase analysis in chapters 7 and 8, unless indicated otherwise, excludes special shows.*

DISTRIBUTION OF SHOWCASE VALUES

Regular watchers of *TPIR* know that one showcase is usually considerably more expensive than the other. Although this may appear intentional, it is not inconsistent with those showcases having been randomly selected from the same normal probability distribution. Such random draws would often generate a pronounced difference between the lower and higher priced showcases. Figure 7.3 shows the distribution of showcase values. The normal curve (i.e., Bell curve) is a reasonably good representation, except for high outliers, a larger-than-expected cluster in the mid-high $20,000 range, and the absence of sub-$20,000 showcases.

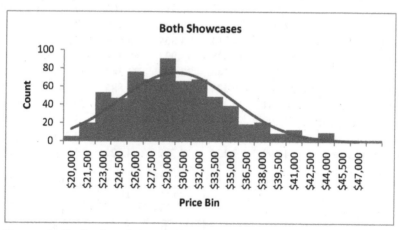

Figure 7.3 Distribution of Showcase Values

*Special episodes—season 47: Premier Episode (9/17/18), Breast Cancer Awareness (10/1/18), Big Money Week (10/8–10/12/18), Veterans Day (11/9/18), pre-Christmas (12/24/18), First Responders (12/28/18), New Year's Eve (12/31/18), Valentine's Day (2/14/19), Preschoolers (4/22/19), Dream Car Week (5/27–5/31/19); season 48: Veterans Day (11/11/19), pre-Christmas (12/22/19, 12/23/19 evening, 12/24/19 day), Valentine's Day (2/14/20), RuPaul evening episode (5/11/20). Big Money Week progressive jackpot for Showcase winner excluded; showcase values otherwise normal. Dream Car Week (2/17/20–2/21/20) not excluded, since showcase values were typical.

SECOND SHOWCASE UNDERPERFORMANCE

Perhaps no takeaway is as surprising as the first showcase win rate eclipsing the second showcase win rate. The bidder on the second showcase should arguably do better, not worse, than the bidder on the first showcase. The second showcase allows one to adjust his bid based on whether he thinks the bidder on the first showcase has bid too aggressively or too conservatively. *Yet the first showcase had the winning bid 52.7% of the time versus 41.3% for the second showcase*, or 56.1% of the time excluding double overbids. I have boxed these results in Figure 7.2.

Why would the second showcase underperform? This goes against all game theoretic and logical intuition. Were second showcases harder to price? Very unlikely. The average price was $30,168 on the second showcase versus $29,515 on the first showcase, hardly a large enough difference to trigger larger bidding errors. *But the second showcase did trigger a much larger number of overbids*, losing by way of an overbid 29.5% of the time versus 18.7% for the first showcase. This is highlighted at the bottom of Figure 7.2. Notably, this 10.8 percentage point difference in overbids almost perfectly matched the 11.4-point win rate advantage in the first showcase versus the second showcase (52.7% versus 41.3%). Bidders on the second showcase bid more aggressively, with a median $2,422 underbid, versus a median $3,400 underbid on the first showcase. Such aggressive bidding often backfired, leading to a higher number of overbids. When neither bidder overbid, the bidder on the second showcase still won only 49.5% of the time. Even in the absence of overbidding, his aggressive bidding did not deliver superior results.[†]

Although the median underbid on the second showcase was only $1,000 less than the median underbid on the first showcase, that lower cushion was problematic on second showcases with lower values. *Too many contestants overbid on lower priced showcases presenting second, driving underperformance on the second showcase.* I return to this subject after reviewing the related topic of "winner" underperformance.

"WINNER" UNDERPERFORMANCE

Equally surprising in the Showcase was underperformance of the bidder in the "winner" position. The "winner" won the Showcase 42.8% of the time versus 51.2% for the "runner-up," performing almost as poorly as the bidder on the

---

[†]49.5% is calculated as follows: The second showcase was won 137 times, 95 net of 42 wins from overbids on the first showcase. These 95 wins were 50% of the 192 Showcases excluding overbids: 332 games less 42 overbids on the first showcase, 78 overbids on the second showcase, and 20 double overbids.

second showcase. This is no accident. The "winner" passed on the first showcase and bid on the second showcase 66% of the time, which should have led to expected underperformance. If the "winner" had won the first showcase 52.7% of the time and the second showcase 41.3% of the time, as boxed in Figure 7.2, his win rate would have been (66.0% × 41.3%) + (34.0% × 52.6%) = 45.2%.[‡]

The "winner's" 42.8% win rate was even worse than the 45.2% expected based on his first/second showcase split. This was mainly a function of chance, as some close bids by the "winner" on the second showcase were outdone by even closer bids by the "runner-up" on the first showcase. The "winner" also delivered some bids that were so low or so high so as to not be credible; for example, a $38,000 bid on a $25,000 showcase.

LOWER PRICED VERSUS HIGHER PRICED SHOWCASES: BIDDING DYNAMICS

The poor win rate of contestants bidding on the second showcase was astonishing, and the poor win rate of the "winner" was also quite notable. The underperformance of the higher priced showcase, however, was marginal. As seen in Figure 7.4, the higher priced showcase had the winning bid 46.1% of

	Lower Priced	Higher Priced	1st - Lower Priced	2nd - Lower Priced	1st - Higher Priced	2nd - Higher Priced
		All Showcase Observations Less 24 Special Shows				
Median	$26,507	$31,928	$26,911	$25,915	$31,277	$32,473
Standard Deviation	3,743	5,006	3,845	3,584	4,709	5,195
Average Underbid	1,631	4,977	2,350	746	5,487	4,374
Median Underbid	1,463	4,583	1,815	493	5,148	4,360
Stdev Underbid	3,982	4,760	3,907	3,875	4,984	4,531
Win	159	153	94	65	81	72
Win %	47.9%	46.1%	52.5%	42.5%	52.9%	40.2%
Loss: Other Player	153	159	72	81	65	94
Loss %	46.1%	47.9%	40.2%	52.9%	42.5%	52.5%
Loss: Double Over	20	20	13	7	7	13
% Double Over	6.0%	6.0%	7.3%	4.6%	4.6%	7.3%
Over	116	44	47	69	15	29
% Over	34.9%	13.3%	26.3%	45.1%	9.8%	16.2%
% Losses Over	67.1%	24.6%	55.3%	78.4%	20.8%	27.1%

Figure 7.4 Showcase Statistical Summary: Lower Priced versus Higher Priced and First versus Second

---

[‡]Noncredible bids are included in win rate statistics, but not in calculating the average underbid (i.e., "cushion").

the time versus 47.9% for the lower priced showcase. The lower priced showcase averaged $27,000 in value versus $32,700 for the higher priced showcase, suggesting smaller bidding errors for the lower priced showcase. After all, a bid that's 10% off on a $40,000 showcase is $4,000 off, whereas a bid that's 10% off on a $20,000 showcase is only $2,000 off. This lower variation in bidding error was observed, with a $3,700 standard deviation of bidding error on the lower priced showcase versus $5,000 on the higher priced showcase. However, this benefit of lower bidding error was offset by frequent overbids on the lower priced showcase, a glaring 34.9% of the time.

For the lower priced showcase, average bids including overbids were only $1,630, below the average showcase price, or 6% of the average showcase value. But the standard deviation of the bidding errors was $4,000, or 15% of the showcase value, explaining why many contestants ended up overbidding when averaging a $1,630 "cushion." This lack of cushion speaks to why 35% of bids on the lower priced showcase were overbids, and 67% of losses on the lower priced showcase were overbids. When neither player overbid, the bidder on the lower priced showcase prevailed 70% of the time.[§]

In contrast, bids on the higher priced showcase averaged $5,000 below the showcase price, or 15.4% of the showcase value. That larger cushion, combined with a $4,750 standard deviation in the bidding error (14.9% of the showcase value) explains why only 13% of bids were overbids. If anything, bidders on the higher priced showcase were too conservative. But that conservative bidding was actually beneficial because of the propensity for bidders on the lower priced showcase to overbid.

Given the overbidding trends, I divided up lower priced and higher priced showcases based on whether they presented first or second. This is also shown in Figure 7.4. When the lower priced showcase presented second, it resulted in an overbid 45% of the time. Median bids were only $493 below the showcase value—far too close for comfort and hardly optimal.

So were contestants tempted to overbid on lower priced second showcases in response to higher bids on the first showcase? Hardly. If anything, the opposite was true. Higher bids on a higher priced first showcase (>$27,000) preceded overbids 35% of the time on the second showcase, whereas lower bids on a higher priced first showcase (<$27,000) preceded overbids on the second showcase 53% of the time. Interestingly, higher priced showcases presenting second also resulted in more overbids—16% of bids versus 10% of bids when

---

[§]70% is calculated as follows: The lower priced showcase won 159 times, 135 excluding 24 wins from overbids on the higher priced showcase. These 135 wins were 70% of the 192 Showcases without overbids: 332 games less 96 overbids on the lower priced showcase, 24 overbids on the higher priced showcase, and 20 double overs.

presented first. This further substantiates bidders bidding more aggressively when bidding second, regardless of whether bidding on the lower or higher priced showcase.

It is reasonable to conclude that players get anxious after observing the first show-case bid, fearing that the showcase bid will be close and leading them to bid too aggres-sively on the second showcase. For the lower priced showcase, this turned a robust average/median $2,350/$1,815 cushion on the first showcase into a far too close for comfort average/median $746/$493 cushion on the second showcase. For higher priced showcases, where underbidding was pronounced, an average/median $5,487/$5,148 cushion on the first showcase became a less conservative average/median cushion of $4,374/$4,360 on the second showcase. The anxiety of bidding second actually helped offset some of the propensity to underbid on the higher priced showcase. As a result, bids on higher priced showcases presenting second prevailed with nearly the same frequency as bids on lower priced showcases presenting second (40.2% win rate vs. 42.5%).

Although characterizing showcases as lower priced and higher priced is an ex–post facto characterization, a similar pattern of overbidding holds if one looks at overbidding on less expensive showcases. First showcases priced less than $28,000 (whether the lower priced or higher priced showcase) resulted in overbids 50 out of 135 times (37%), whereas second showcases priced less than $28,000 resulted in overbids 67 out of 134 times (50%).

Let's revisit the frequency of overbids on lower priced showcases presenting second. Going over 45.1% of the time means a maximum win rate of 54.9%. Even though the lower priced showcase presenting second won 77.4% of the time when not going over, that only allowed for a 77.4% × 54.9% = 42.5% win rate.

For a dose of irony, compare the 45.1% overbid rate on lower priced showcases presenting second to the 27.5% overbid rate in Contestant's Row when including an estimate for unaired overbids. As discussed in chapter 4, bidders in Contestant's Row would have benefitted by bidding aggressively and going over more than 40% of the time. In contrast, bidders on the lower priced showcase would have benefitted from going over about 26% of the time, as they did when the lower priced showcase was presented first.

Overcompensating due to the anxiety of going second can detract from performance off the set of *TPIR*. Most contests are back and forth. The second player knows the score he or she has to beat, whether gymnastics competitions or tie-breaking soccer penalty kicks in elimination rounds. Professor Palacios-Huerta of the London School of Economics studied more than ten thousand penalty kicks in major national and international competitions held between 1970 and 2012. He found that the team that took the first penalty kick won 60% of the time, whereas the team that took the second kick won 40% of the time, evidence of anxiety from going second.

Occasionally the second player does not know how the first player has scored. In a court case, the defense presents its closing arguments second. Although the defense can rebut the prosecution's arguments one by one, it risks overcompensating and pushing back against arguments made by the prosecution that have little sway with the jury. Such arguments may distract the jury from focusing on the defense's rebuttals to what the jury feels are the prosecutor's core suppositions. In effect, the defense risks overbidding. The same can happen in a job interview, particularly if one knows the hiring decision comes down to a final round interview among two or three candidates. Here one need not interview last to be anxious about how the other candidates have performed. The goal of a final round interview in many ways is to be the closest to what the employer is looking for without going over. "Going over" can take the form of being overqualified, being perceived as a threat by one's future boss, or being perceived as too hard-charging.

SECOND SHOWCASE UNDERPERFORMANCE: PRESCRIPTIVE STRATEGIES

So what is the right level of aggressiveness when bidding on lower and higher priced showcases? Conceptually this is straightforward. A contestant need only compare his probability of going over and automatically losing with his probability of losing if he does not go over. If the probability of going over is high, the bidder is likely bidding too aggressively. If the probability of overbidding is low but the probability of winning when not overbidding is also low, the contestant is not bidding aggressively (i.e., high) enough. In this case, bid more aggressively, go over a bit more, but win a greater share of Showcases when not going over.

Of course, contestants bidding on the second showcase did a poor job of following this strategy on lower priced showcases and struggled on higher priced showcases due to general imprecision. *Bidding on the first showcase resulted in better outcomes—both win rate and average showcase earnings—regardless of whether that showcase was lower priced or higher priced.* This explains the surprising underperformance on the second showcase.

So how can contestants take advantage of underperformance on the second showcase, and how can they avoid the anxiety-fueled overbidding bias on the second showcase? The answers are straightforward.

- Bid on the first showcase if you are the "winner." Enough said.
- If bidding on a lower priced second showcase, bid as if that showcase were presented first. Try to correct for the anxiety associated with bidding second, and bid with a greater margin of safety.

Of course, these prescriptions are the opposite of what was observed in seasons 47 and 48. The "winner" bid on the second showcase 66% of the time, won only 43% of the time, and pocketed average showcase earnings of $12,250 versus $13,850 for the "runner-up." The "runner-up" bid on the less expensive showcase 56% of the time, but won 51% of the time, which more than offset a slightly lower average showcase value. The root cause of the differential performance was overbidding on the second showcase.

Remember, a contestant bidding second should fare no worse than if bidding on a similarly priced showcase presenting first, whether it is the lower priced or higher priced showcase. In theory he should perform better.

To gauge how conservatively contestants should bid on the second showcase (i.e., how far under the retail price they should target), one could shift the bids on the second showcase by a fixed percentage or dollar amount and see if the win rate increases. But this has been done for us—just look at how successfully contestants bid when the lower priced showcase is presented first. As seen in Figure 7.4, when the lower priced showcase presented first, the bidder had an average/median underbid of $2,350/$1,815. When the lower priced showcase presented second, the bidder had an average/median underbid of $746/$493. *Had a bidder on the lower priced showcase presenting second bid 6% lower so that his average underbid was $2,300 below the showcase price, he would have won 54.9% of the time, not 42.5% of the time.*

The second bidder needs to *overcome his cognitive biases and bid as if he were bidding first, which would result in him lowering his bid by approximately $1,500 for an identical showcase.* Easier said than done, as the contestant bidding second is likely not aware of his bias to bid too aggressively. Still, he can tell himself to mentally block out the bid on the first showcase. Or, if bidding on a less expensive second showcase, he can remind himself that the first showcase was likely higher priced with a bid that was likely meaningfully too low. Average/median bids for the first showcase, if higher priced, were $5,487/$5,148 below the retail price. With showcases never pricing below $20,000, and a median lower priced showcase of $26,500, it should not be difficult to construct a competitive mid-$20,000 bid for a lower priced showcase presenting second. The effect should be a win rate of 52.5%, consistent with the lower priced showcase presenting first, and versus the 42.5% win rate observed.

Fortunately for the "winner," he can always choose to bid on the first showcase, avoiding the mental challenges of having to bid second. Of course, in some cases the first showcase may not be to his liking, or the price may be difficult to estimate. Otherwise, he should bid on the first showcase and benefit from the overbidding on the second showcase.

But why should this strategy work prospectively? Because there is no reason for the psychological errors by the second bidder to disappear, outside of the lucky souls who read this book. I have every reason to believe that by bidding first every time, the "winner" would have delivered about 170 wins and 142 losses instead of 142 wins and 170 losses.

ADDITIONAL SHOWCASE OBSERVATIONS TIED TO PRICE AND BID ORDER

A few other observations of interest from the Showcase tied to showcase price and bid order:

- Notwithstanding the 35% frequency of overbids on the lower priced showcase, an overbid on the higher priced showcase was usually the gating factor to a double overbid, because the higher priced showcase went over only 13% of the time.
- The lower priced showcase was presented first 179 times (54%) and second 153 times (46%).
- The most expensive showcases, even excluding the special shows, were typically presented second. The higher priced showcase averaged $32,100 when presented first and $33,300 when presented second.
- Blowout $40,000-plus showcases were offered thirty-three times, and fifty-three times including special shows, with sixteen out of twenty-four special shows offering one blowout showcase and four offering two blowout showcases. Blowout showcases were won only about 30% of the time.
- The average bid on a higher priced showcase presenting first was $5,500 below the showcase value. This was not negative, though, given the frequency of overbids on lower priced showcases.

Key Tip

As the "winner," bid on the first showcase to benefit from frequent overbids on the second showcase. Make exceptions for personal preference, but do not fall prey to "the grass is greener" mentality. If bidding on what you think is the lower priced showcase, target a bid about $2,000 below the retail price of the showcase. This may not be a challenge if bidding first but beware of bidding too aggressively and overbidding if bidding second.

8

Showcase—Other Observations

Other topics of interest in the Showcase include the effect of cars and boats on showcase values and winnings, how the winner chose which showcase to bid on, tips that can help contestants identify the higher priced showcase, and some pointers for sizing up showcase values.

A few key observations:

- In 356 Showcase contests, 462 cars were offered, or 1.30 cars per Showcase contest, and in both showcases 30% of the time. Boats were offered in 92 of 356 showcase contests (26%) and never on both showcases. Boats were generally undesired and had unfamiliar values, leading to poor outcomes. Showcases with boats were passed frequently when part of the first showcase.
- Bidding on showcases without a car—despite allowing players to average bidding errors across prizes—led to a lower win rate. Cars seemed to be more known entities than prizes like vacations and furniture.
- The "winner" bid on the higher priced showcase 56% of the time and the second showcase 66% of the time. He likely could have bid on the higher priced showcase 60 to 65% of the time if so motivated.

KEY SHOWCASE ITEM ANALYSIS

Cars and boats stand out because they are a disproportionate share of the value of the showcase, even though showcases typically feature three prizes or "prize families" (e.g., a living room consisting of a sofa, recliner, and coffee table). The average boat or car is valued at about $20,000 and comprises two-thirds of the average $30,000 showcase. Making an accurate bid is driven in large part by accurately assessing the value of the car or boat.

Across both showcases, *excluding special shows*, 1.28 cars and 0.26 boats were offered on average. Of the lower priced showcases (N = 332), 66% offered a car or a boat, with 182 offering a car (55%) and 38 (11%) offering a boat. Of the higher priced showcases, 88% featured a car or a boat, with 244 cars (73%) and 50 boats (15%). One hundred and twelve lower priced showcases (34%) featured neither a car nor a boat, compared to 39 (12%) higher priced showcases. Cars appeared in both showcases close to one-third of the time.

Only once during seasons 47 and 48 were a car and boat proffered in the same showcase, which was won by the other contestant in a double showcase win. And only once, on the season 47 Valentine's Day special, were two cars proffered, namely to a male and female paired for a blind date, which they won. When higher priced showcases featured neither a car nor a boat, they instead offered prizes like expensive vacations; for example, a trip around the world with three stops. As best as I could identify, cars valued at more than $30,000 were relatively uncommon, appearing only about a dozen times excluding special shows and two dozen times including special shows. Much more common was the typical $18,000 to $22,000 car.

A car or a boat prevents the contestant from averaging offsetting errors across similarly priced items in a showcase (i.e., erring low on the price of one prize and erring high on the price of another prize). Nonetheless, contestants bidding on lower priced showcases did better with a car in their showcase.* They won 51.1% of the time with a car versus 44.0% of the time without a car and 43.8% of the time without a car or a boat. The standard deviation of the miss was actually a touch lower ($3,900) with the car than without ($4,000), suggesting no benefit from estimation errors canceling out on showcases with a few moderately priced prizes. Contestants were also more likely to bid without a margin of safety when there was no car or boat in the showcase, going over with greater frequency.

Contestants won only thirty-five out of eighty-seven times (40%) when bidding on a showcase with a boat, worse than any of my other data cuts. Boats hurt contestant's chances the most when they were expensive, and contestants dramatically underestimated their cost. On lower priced showcases, boats did not lower the win rate, perhaps because a tendency to underestimate the boat's price neutralized a tendency to bid too aggressively on lower priced showcases. Nor did contestants want boats. Forty boats were offered in the first showcase, and the "winner" passed thirty-six out of forty times.

---

*Only thirty-nine higher priced showcases had no car or boat, making the sample size too small to effectively analyze.

"WINNER" TARGETING SECOND SHOWCASE

In chapter 7 I discussed the pitfalls of bidding on the second showcase. So why did the "winner" pass on the first showcase in 219 out of 332 episodes—nearly two-thirds of the time? And why did he pass more than half the time when bidding on showcases priced more than $33,000 (corresponding to top quartile showcases)? If the "winner" had a sixth sense as to which showcase was priced higher, it was a weak sixth sense. He bid on the higher priced showcase only 56% of the time. Still, I believe the "winner" would have gravitated to the higher priced showcase more but for the following considerations:

1. A tendency to want to see the second showcase (i.e., "grass is greener on the other side" mentality).
2. Avoidance of boats when part of the first showcase.

The data supports the "grass is greener on the other side" mentality. When the first showcase was priced more than $33,000, the "winner" bid on it only thirty-five out of seventy-one times (49%). Only when valued at more than $36,000 did the "winner" lean toward the first showcase—and only modestly so—bidding sixteen out of thirty times (53%). I would have expected more bidding on pricier first showcases by the "winner," even if he could not identify which showcases were pricier with certainty.

I believe the "winner" could have bid on the higher priced showcase 60 to 65% of the time had that been his sole focus. Why is 60 to 65% a good bogey? If the "winner" knew the value of the first showcase with certainty and the distribution of showcase values, including its percentile ranking, he would have had 75% certainty as to whether it was the higher priced showcase.[†] Of course, the "winner" cannot deliver on this 75% outcome because he does not know the price of the first showcase with certainty. Nor is the normal distribution (i.e., Bell curve) a perfect representation of how showcases are priced. Most importantly, there is personal preference at play; for example, an aversion to boats and a desire to see if the other showcase is better.

[†]Suppose the first showcase is in the sixtieth percentile. Then there is a 60% chance it is higher priced. If it is in the thirtieth percentile, there is a 30% chance it is higher priced, meaning 70% certainty on its ranking. To figure out the average across all possible scenarios, we need to integrate $f(x) = x$ from the fiftieth to one hundredth percentile and multiply by two:

$$2 \times \int_{0.5}^{1.0} x\,dx = 2 \times \left[\frac{x^2}{2}\right]_{0.5}^{1.0} = 2 \times [1/2 - 1/8] = 2 \times 3/8 = 0.75, \text{ or } 75\%.$$ The 75% assumes showcases are drawn from the *same* distribution.

TARGETING THE HIGHER PRICED
SHOWCASE AND BIDDING TIPS

Targeting the higher priced showcase is not a winning strategy, personal preferences aside. Rather, given the propensity to overbid on lower priced showcases presenting second and the lower likelihood of winning on higher priced showcases presenting second, the "winner" should bid on the first showcase.

But what if the "winner" wants to bid on the higher priced showcase? What heuristics can he follow to identify the expensive showcase and craft an effective bid? Here are a few considerations:

- First showcases without a car or a boat were lower priced 71% of the time.
- Non-standard cars almost always result in a higher priced showcase.
- Showcases with pontoon boats or sailboats averaged $33,000, or twenty-fifth percentile. Showcases with generic speed boats averaged $30,000 and were the higher priced showcase about half the time. Although I do not know a boat's contribution to the showcase, my best guess is that typical pontoon boats run $21,000 to $22,000 and typical speed boats run $18,000 to $19,000.
- Cars typically have values within a few thousand dollars of $20,000 on *The Price Is Right*. A player can approximate these values by edging up to $21,000 or $22,000 for leading car models, like a Toyota Corolla or Honda Civic, and edging down to $17,000 or $18,000 for discount car models, like a Hyundai Accent or Ford Fiesta. Models like Volkswagen Jettas or Hyundai Elantras tend to run around $20,000.
- For vacations, an intercontinental trip might cost $11,000 to $12,000, and more with extras, whereas a continental six-night trip might run $8,000 to $9,000, and perhaps $1,000 or $2,000 less if only five nights.

The aforementioned considerations are useful in approximating the value of a showcase's main prize before approximating the value of the other prizes and applying a margin of safety in one's bid. If the contestant follows these guidelines and estimates a showcase value materially above $30,000 (pre-"margin of safety"), he is likely looking at the higher priced showcase.

Key Tip

When bidding on one's showcase, one expensive prize usually drives the lion's share of the showcase value. Consider standard price ranges for cars, boats, and vacations when mentally tabulating an estimated value for one's showcase, and then apply an appropriate margin of safety.

9

Flow of the Show

Before launching into a discussion of the pricing games, it is worthwhile to discuss contestant performance holistically across the show—how much money was awarded across the show and how contestants in different bidding positions performed, starting in Contestant's Row, and carrying through into the pricing games and the Showcase. I have chosen to refer to the person who bids last in the first bidding round as Contestant 1, since he has the best chance of winning. In effect, Contestants 1–4 are Bidders 4–1 in the first bidding round. Contestant 5 is the contestant called down from the audience for the second bidding round, and so forth through Contestant 9, who is called down for the sixth bidding round.

Figure 9.1 tracks contestant performance with some interesting observations. First, there is very little difference in the frequency with which Contestants 1–4 made it out of Contestant's Row, each making it on stage more than 80% of the time. Having six "at bats" was the big equalizer for the first four contestants. Interestingly, Contestant 1, despite the benefit of bidding last in the first bidding round, got on stage no more frequently than Contestant 4, who performed very well in the second half of the show. Perhaps bidding first

| | Contestant's Row | | Pricing Game | | | | | | | | Showcase | | | | |
Contestant Number	Bid Wins	To Pricing Game	1	2	3	4	5	6	Total	Game Win %	In Showcase No	In Showcase Yes	Showcase Wins	Showcase Win Rate	Showcase Wins (%)
1	316	89%	134	52	50	34	24	22	316	50%	197	119	69	58%	21%
2	295	83%	83	89	45	33	27	18	295	47%	199	96	42	44%	13%
3	295	83%	78	77	58	31	40	11	295	46%	208	87	37	43%	11%
4	314	88%	61	78	69	54	29	23	314	50%	210	104	48	46%	14%
5	270	76%	0	60	61	67	53	29	270	49%	168	102	46	45%	14%
6	244	69%	0	0	73	79	56	36	244	54%	175	69	27	39%	8%
7	197	55%	0	0	0	58	71	68	197	52%	134	63	31	49%	9%
8	134	38%	0	0	0	0	56	78	134	44%	87	47	17	36%	5%
9	71	20%	0	0	0	0	0	71	71	48%	46	25	18	72%	5%
	2,136	67%	356	356	356	356	356	356	2,136	49%	1,424	712	335	47%	100%

Figure 9.1 Contestant Performance across *The Price Is Right*

in the first round forced Contestant 4 to bid what he thought versus becoming over-influenced by other contestants, which helped Contestant 4 in later rounds when he bid third or fourth.

Each contestant reached the Showcase about one-third of the time when they got out of Contestant's Row, as one would expect. Contestant 1 performed a bit better, reaching the Showcase 38% of the time (119 times versus 316 pricing games played) and Contestant 6 a bit worse, reaching the Showcase only 28% of the time (69 times versus 244 pricing games played). This was likely due to chance. Performance in the Showcase across contestants raises some interesting questions. Contestant 1 won his showcase 58% of the time (69 out of 119), which should occur with only 2% likelihood. Perhaps getting out of Contestant's Row early helped his confidence in the Showcase. Contestant 9 won his showcase 72% of the time (18 out of 25), also only 2% likely through random chance. Perhaps having only one chance to get out of Contestant's Row made him more focused in the showcase. Contestants 2–6 and Contestant 8 modestly underperformed the average 47% win rate in the Showcase.

Let's dig a bit deeper. Contestants 1 and 9 went over only 17% (20 out of 119) and 16% (4 out of 25) of the time, respectively, on their showcases, compared to 25% for contestants 2–8. Of showcase wins by Contestant 9 (10 out of 18), 56% had him within $2,500 of the showcase's retail price. Four of those ten wins were within $1,000, notable precision given Contestant 9's low overbid rate. Of the showcase wins (32 out of the 69) by Contestant 1, 46% had him within $2,500 of the retail price, also notwithstanding infrequent overbids. In contrast, only 41% of the wins by Contestants 2–8 (101 out of 248) had them within $2,500 of the showcase's retail price, despite more overbidding. Though hard to link cause and effect, I think there is good reason for Contestants 1 and 9 to have more confidence and focus during the Showcase.

Figure 9.2 shows prize money won per contestant. More than $27 million of prize value was won during seasons 47 and 48, with Contestant 1 taking home $5.1 million, or 19% of the total. On the opposite end of the spectrum, Contestant 9 took home only $1.1 million, or 4% of the total. The average winnings per contestant upon making it on stage were higher for Contestant 1, a function of him often playing the first pricing game, which had more lucrative prizes.

The pricing games and the Showcase were the most lucrative parts of the show, as seen in Figure 9.3. On average, contestants won $76,300 per show, of which $37,200 (49%) was in pricing games and $28,700 (38%) was in the Showcase. Contestant's Row added $8,850 per show (12%), and the Showcase Showdown averaged $1,500, reflecting a $1 total achieved by about 8% of spinners and the occasional $10,000 or $25,000 win on one's bonus spin. The Showcase winner on average won 50% of the prize value awarded per show.

Contestant Number	Total Prize Winnings (356 shows)	Average Winnings per Show	Total Pricing Games	Average Winnings If On Stage*
1	$5,101,959	$14,331	316	$16,145
2	3,596,024	10,101	295	12,190
3	3,316,026	9,315	295	11,241
4	3,880,184	10,899	314	12,357
5	3,408,402	9,574	270	12,624
6	2,678,946	7,525	244	10,979
7	2,611,110	7,335	197	13,254
8	1,520,668	4,272	134	11,348
9	1,055,790	2,966	71	14,870
	$27,169,110	$8,480	2,136	$12,720

*Calculated as (Average Winnings Per Show * 356) / Total Pricing Games

Figure 9.2 Contestant Winnings

Season	Total Winnings	Contestant's Row*	Pricing Game Win & Other	Showcase Showdown	Showcase	Showcase Winner	All Other Contestants
47	$75,321	$8,634	$36,345	$1,926	$28,417	$36,367	38,954
48	$77,445	$9,123	$38,179	$1,090	$29,054	$39,669	37,777
Total	$76,318	$8,861	$37,205	$1,534	$28,716	$37,916	38,402

*Includes $500 bonuses from bidding the exact retail price, as occurred in 10% (71 out of 712) bidding rounds

Figure 9.3 Per-Show Winnings

Most shows had contestants winning $50,000 to $100,000 in total, with 71% of shows (252 out of 356) in this range. Positive outliers generally reflected strong pricing game performance, whether because prize values were boosted by special shows (e.g., Big Money Week, Dream Car Week) or because five or six contestants prevailed in their pricing games.

When there was little prize money awarded, it often coincided with no contestant winning his showcase (i.e., a double over) or few contestants winning their pricing games. Still, *The Price Is Right* (*TPIR*) gave away less than $40,000 in only sixteen shows and less than $30,000 in only five shows.

Excluding the 24 special shows, 16 of which had winnings in excess of $100,000, only 39 out of 332 shows had winnings in excess of $100,000. Still, average show winnings of $76,300, or $12,700 per contestant making it out of Contestant's Row, highlights how lucrative *TPIR* can be for contestants. Finally, note that total winnings per show in season 48 increased nearly 3% to $77,350, from $75,350 in season 47, driven by increases in the value of items up for bids in Contestant's Row and the value of prizes in the pricing games. However, the contours of the show remained mostly the same.

The rest of this book analyzes the seventy-seven *TPIR* games, going into more detail where analytically or strategically relevant. Following the pricing game introduction in chapter 10, feel free to read chapters 11 through 21 in any order to learn about the different categories of games.

Please note that the pricing game recommendations are based on prevailing price levels during seasons 47 and 48, the latter culminating in June 2020. Inflation in prize values and inflation in the value of small items used in pricing games may require fine-tuning of some of the recommendations.

Part II

Individual Pricing Games of Chance: Strategy and Behavioral Shortfalls

10

Pricing Games—Overview and
Role of Heuristics

A contestant who gets out of Contestant's Row plays one of seventy-seven pricing games. Prizes can vary from a luxury car on the high end to a set of prizes valued slightly more than $5,000 on the low end. Most pricing games are played for a car valued between $17,000 and $23,000, cash between $10,000 and 20,000, a vacation valued between $8,000 and $12,000, or a set of prizes totaling between $5,000 and $10,000.

A representative show has two car games, one game for a vacation prize, one game for a non-vacation prize, up to one game for cash, and one to two multi-prize games. Some games are designed for car play, and some games are designed for cash play. Simple games in which one guesses the price of a single prize are played for vacations and a variety of non-car prizes. Certain games are set up for two, three, or four prizes. Grocery games can be played for a vacation, a non-vacation prize, or a set of prizes. In some games, contestants guess prices for small items (typically $200 or less) in order to win a larger prize.

I define ten categories of games in parallel with the above discussion and classify each pricing game. The goal is not equally sized groups, but groups with a similar style of play and/or prize type (see Figure 10.1).

I define a win as earnings in excess of $5,000. Never did a pricing game feature total prize value of less than $5,000, analogous to the $20,000 minimum prize value of showcases and a $500 minimum prize value in Contestant's Row bidding. Although players typically win the game's main prize, they sometimes exceed $5,000 by winning lesser prizes, which is possible in most cash games and in some car games.

I define the statistic median earnings in order to measure how rewarding pricing games are on an absolute and relative basis. "Median earnings" multiplies median win value by win rate and adds average earnings less than $5,000 when the contestant failed to win his pricing game. In

2-Item	Price Guess	Cash Games	Grocery Games	Car Pricing	Car Plus
Bargain Game	Balance Game	Grand Game	Bullseye	Any Number	Gas Money
Do The Math	Bonkers	Half Off	Check Out	Card Game	Golden Road
Double Cross	Check Game	Hot Seat	Grocery Game	Cover Up	More or Less
Magic Number	Clock Game	It's in the Bag	Hi Lo	Dice Game	Pass the Buck
One Right Price	Coming or Going	Pay the Rent	Now or Then	Five Price Tags	Temptation
Switch	Double Prices	Plinko	Pick-A-Pair	Gridlock	Three Strikes
	Flip Flop	Punch-A-Bunch	Vend-O-Price	Line 'em Up	Triple Play
3-Item	Freeze Frame	Time is Money		Lucky Seven	
Easy as 1-2-3	Pick-A-Number		**Small Item**	Money game	**Other Car**
Make Your Move	Push Over		Bonus Game	One Away	Hole in One
Most Expensive	Range Game		Cliffhanger	Pathfinder	Let 'em Roll
One Wrong Price	Safe Crackers		Secret X	Pocket Change	Master Key
	Side by Side		Shell Game	Stack the Deck	Rat Race
4-Item	Squeeze Play			Switcheroo	Spelling Bee
Danger Price	2 for the Price of 1			Ten Chances	
Race Game				That's Too Much	
Shopping Spree					
Swap Meet					
Take Two					

Figure 10.1

Plinko, players averaged $12,190 in 24 wins, and $1,696 in 23 losses, for $\left(\frac{24}{47} \times \$12,190\right) + \left(\frac{23}{47} \times \$1,696\right) = \$7,054$ in median earnings. I rank this self-created metric across all pricing games and categories.

Median earnings approximate the expected value of playing a pricing game when excluding abnormally large prizes, as it is less skewed by occasional big prizes, such as those offered during Dream Car Week or Big Money Week. I use median win value instead of median prize value to calculate median earnings so as to capture actual contestant winnings, which are often not the full prize value. Also, *The Price Is Right* (TPIR) frequently makes more expensive prizes harder to win.[*]

During seasons 47 and 48, 2,136 pricing games were played across 356 episodes, with all but five *TPIR* games.[†] Of the 2,131 *TPIR* games played, 1,045 were won, for a 49% win rate, with median prize value slightly more than $15,000. The average win value of $12,335 and median win value of $11,227 were lower as the contestant could win more than $5,000 without winning the full prize.[‡] This included *Pay the Rent*, in which a $10,000 victory was far more common than a $100,000 victory, and *Temptation*, in which players often won

---

[*]Four exceptions in how I calculate median earnings. In (1) *Grocery Game*, I use median prize value instead of median win value, as the seven wins were much lower in value than the median prize value, likely a function of chance. In (2) *Now or Then*, (3) *Pay the Rent*, and (4) *Time Is Money*, I use average win value. Median win value is less useful in these games since there are wide-ranging win scenarios or a cluster of lower value wins.

[†]This excludes five *Let's Make a Deal* games played during Mash Up Week in season 48.

[‡]The average prize value and median prize value across all pricing games were weighted by the number of games played, and the average and median win values by the number of wins.

in excess of $5,000 by taking the prizes instead of playing for the car and the prizes. Median win value fell below average win value on account of expensive prices, often offered on special shows.

Median earnings were $5,720 across all pricing games, excluding *Golden Road*, *Three Strikes*, and *Triple Play*. Those three games were played very infrequently for large prizes but with low win rates. My best estimate of the "middle amount" that contestants earned across pricing games, which is not skewed upward by special shows with very large prize values, is therefore $5,720. It includes players' median winnings ($5,576) and their average earnings ($144) when they failed to win the game but still took home some prize value.

The five most played pricing games—each played more than fifty times—during seasons 47 and 48 were *Squeeze Play* (57), *Lucky Seven* (55), *Double Prices* (53), *Flip Flop* (52), and *Bargain Game* (51). One can see from Figure 10.2 that pricing games generally fell in three categories in terms of frequency:

- 18 frequently played games, each played more than 40 times for a total of 866 games (41%);
- 28 moderately played games, each played 20 to 40 times for a total of 790 games (37%);
- 31 infrequently played games, each played less than 20 times for a total of 475 games (22%).

Figure 10.3 aggregates the statistics in Figure 10.2 by category of game. Car games and price guess games like *Flip Flop* and *Side by Side* are the core of *TPIR*, accounting for 32% and 25% of pricing games during seasons 47 and 48, respectively. Cash games were played about twice every three episodes. Grocery and small item pricing games, on a combined basis, were also played about twice every three episodes.

Contestants want to play for cars on *TPIR* and rightly so. Car games are both exciting and lucrative. Average prize value and median win value both approximate $20,000 in car games, compensating for a modest 36% win rate. On average, contestants earned about $240 when they did not win a car game, bringing median earnings to slightly more than $7,200. This surpassed any other pricing game category. Among car games, "other" car games had the highest win rate, 45%, largely owing to success in *Hole in One (or Two)*. "Car plus" games, typically played for a luxury car, had a low 30% win rate, which negated the high prize value. That win rate would have been even lower excluding multiple non-car wins in *Golden Road*, in which the contestant won prizes valued north of $5,000.

In contrast, non-car games had a modest $8,800 median win value and a higher 55% win rate. Among categories of non-car games, small item pricing

Game	Inaugural Year	Type	# Times Played	# Times Won	# Times Lost	Win Rate	Average Prize	Median Prize	Average Win	Median Win	Min Win	Max Win	Median Earnings	Earnings Rank
Any Number	1972	Car Pricing	42	12	30	29%	$ 22,432	$ 21,446	$ 21,048	$ 21,559	$ 16,257	$ 24,841	$ 6,453	26
Balance Game	2006	Price Guess	36	15	21	42%	$ 11,193	$ 10,304	$ 11,093	$ 9,938	$ 6,318	$ 18,215	$ 4,141	55
Bargain Game	1980	2-Item	51	41	10	80%	$ 8,624	$ 7,355	$ 8,744	$ 7,355	$ 5,278	$ 32,548	$ 5,913	31
Bonkers	2001	Price Guess	17	9	8	53%	$ 7,606	$ 7,593	$ 7,433	$ 7,549	$ 5,942	$ 9,349	$ 3,997	58
Bonus Game	1972	Small Prize	16	13	3	81%	$ 10,175	$ 9,819	$ 10,070	$ 9,458	$ 6,101	$ 18,264	$ 7,717	15
Bullseye	1976	Grocery	23	21	2	91%	$ 11,494	$ 12,222	$ 11,552	$ 12,222	$ 5,000	$ 25,000	$ 11,159	3
Card Game	1974	Car Pricing	31	11	20	35%	$ 22,866	$ 22,885	$ 22,052	$ 22,070	$ 21,111	$ 23,496	$ 7,831	14
Check Game	1981	Price Guess	16	6	10	38%	$ 7,875	$ 7,500	$ 8,094	$ 8,034	$ 7,550	$ 8,930	$ 3,013	66
Check Out	1982	Grocery	23	6	17	26%	$ 10,937	$ 10,000	$ 12,481	$ 11,532	$ 5,018	$ 19,575	$ 3,008	67
Cliffhanger	1976	Small Prize	31	26	5	84%	$ 12,658	$ 9,305	$ 13,006	$ 9,996	$ 5,197	$ 72,044	$ 8,384	12
Clock Game	1972	Price Guess	13	8	5	62%	$ 13,804	$ 10,420	$ 15,184	$ 10,515	$ 7,369	$ 51,285	$ 6,668	24
Coming or Going	2003	Price Guess	39	26	13	67%	$ 7,406	$ 7,619	$ 7,564	$ 7,678	$ 6,078	$ 9,056	$ 5,118	39
Cover Up	1993	Car Pricing	39	16	23	41%	$ 20,995	$ 20,814	$ 21,239	$ 20,430	$ 16,425	$ 31,428	$ 8,382	13
Danger Price	1976	Car Pricing	18	5	13	28%	$ 8,767	$ 8,436	$ 9,477	$ 8,539	$ 8,092	$ 12,656	$ 2,372	74
Dice Game	1976	Car Pricing	33	16	17	48%	$ 19,492	$ 21,225	$ 19,991	$ 21,308	$ 15,345	$ 24,345	$ 10,331	7
Do the Math	2013	2-Item	48	31	17	65%	$ 9,129	$ 8,157	$ 9,850	$ 8,778	$ 6,464	$ 16,530	$ 5,669	33
Double Cross	2012	2-Item	23	14	9	61%	$ 7,090	$ 6,485	$ 7,387	$ 6,411	$ 5,548	$ 14,023	$ 3,902	59
Double Prices	1972	Price Guess	53	33	20	62%	$ 9,812	$ 9,498	$ 9,614	$ 9,619	$ 5,338	$ 18,298	$ 5,989	30
Easy as 123	1996	3-Item	19	7	12	37%	$ 7,444	$ 7,262	$ 7,145	$ 6,943	$ 6,334	$ 8,219	$ 2,558	72
Five Price Tags	1972	Car Pricing	19	10	9	53%	$ 25,862	$ 20,516	$ 26,083	$ 20,849	$ 17,388	$ 79,791	$ 11,014	4
Flip Flop	2000	Price Guess	52	31	21	60%	$ 7,780	$ 7,588	$ 7,606	$ 7,565	$ 5,719	$ 9,783	$ 4,510	49
Freeze Frame	1995	Price Guess	35	14	21	40%	$ 7,509	$ 7,429	$ 7,150	$ 6,790	$ 5,598	$ 9,148	$ 2,716	69
Gas Money	2008	Car Plus	20	4	16	20%	$ 30,346	$ 30,358	$ 22,513	$ 26,781	$ 9,000	$ 27,490	$ 5,356	36
Golden Road*	1975	Car Plus	7	5	2	71%	$ 86,638	$ 80,863	$ 20,116	$ 7,721	$ 6,731	$ 70,383	N/A	N/A
Grand Game	1980	Cash	40	12	28	30%	$ 13,000	$ 10,000	$ 10,000	$ 10,000	$ 10,000	$ 10,000	$ 3,125	65
Gridlock	2017	Car Pricing	24	11	13	46%	$ 20,369	$ 20,500	$ 19,975	$ 20,413	$ 15,863	$ 21,938	$ 9,356	10
Grocery Game	1972	Grocery	17	7	10	41%	$ 11,807	$ 10,863	$ 10,136	$ 8,350	$ 5,222	$ 20,940	$ 4,473	50
Half Off	2004	Cash	34	9	25	26%	$ 12,599	$ 11,397	$ 15,464	$ 11,286	$ 10,308	$ 51,594	$ 3,243	64
Hi Lo	1973	Grocery	21	11	10	52%	$ 10,164	$ 8,900	$ 8,927	$ 8,700	$ 5,366	$ 17,530	$ 4,557	45
Hole in One	1977	Other Car	16	11	5	69%	$ 23,203	$ 17,716	$ 17,808	$ 17,315	$ 13,185	$ 23,840	$ 11,904	1
Hot Seat	2016	Cash	24	11	13	46%	$ 23,652	$ 20,314	$ 8,863	$ 10,173	$ 5,140	$ 20,210	$ 4,963	42
It's in the Bag	1997	Cash	25	6	19	24%	$ 18,560	$ 16,000	$ 12,000	$ 12,000	$ 8,000	$ 16,000	$ 3,600	60
Let 'em Roll	1999	Other Car	23	7	16	30%	$ 25,154	$ 18,985	$ 38,596	$ 17,070	$ 5,000	$100,000	$ 6,891	22
Line 'em up	1998	Car Price	23	12	11	52%	$ 19,461	$ 19,561	$ 19,465	$ 19,500	$ 17,561	$ 20,995	$ 10,174	8
Lucky Seven	1973	Car Pricing	55	17	38	31%	$ 20,764	$ 21,492	$ 21,045	$ 21,593	$ 17,298	$ 25,645	$ 6,674	23
Magic Number	1992	2-Item	13	11	2	85%	$ 6,778	$ 6,445	$ 6,913	$ 6,477	$ 5,448	$ 9,262	$ 5,481	35
Make Your Move	1989	3-Item	16	6	10	38%	$ 7,590	$ 7,289	$ 7,346	$ 7,113	$ 5,978	$ 8,749	$ 2,667	70
Master Key	1983	Other Car	16	4	12	25%	$ 23,216	$ 23,198	$ 23,389	$ 23,389	$ 21,924	$ 25,106	$ 6,522	25
Money Game	1972	Car Pricing	46	25	21	54%	$ 20,286	$ 20,616	$ 19,820	$ 19,891	$ 16,352	$ 23,211	$ 10,905	6
More or Less	2007	Car Plus	18	5	13	28%	$ 18,470	$ 24,142	$ 13,681	$ 7,259	$ 6,354	$ 24,611	$ 4,094	56
Most Expensive	1972	3-Item	44	25	19	57%	$ 7,907	$ 7,590	$ 8,079	$ 7,598	$ 6,600	$ 16,227	$ 4,317	52
Now or Then	1980	Grocery	17	15	2	88%	$ 8,920	$ 6,732	$ 8,423	$ 6,732	$ 5,000	$ 14,846	$ 7,432	17
One Away	1984	Car Pricing	32	8	24	25%	$ 23,211	$ 20,903	$ 25,481	$ 21,418	$ 18,257	$ 35,271	$ 5,355	37
One Right Price	1975	Car Pricing	46	25	21	54%	$ 7,497	$ 6,220	$ 7,206	$ 6,245	$ 5,216	$ 12,623	$ 3,394	62
One Wrong Price	1998	3-Item	49	22	27	45%	$ 7,745	$ 7,668	$ 7,847	$ 7,714	$ 6,087	$ 10,204	$ 3,463	61
Pass the Buck	2001	Car Plus	22	9	13	41%	$ 22,262	$ 20,468	$ 13,240	$ 16,195	$ 5,000	$ 23,115	$ 7,443	16
Pathfinder	1987	Car Pricing	24	5	19	21%	$ 20,632	$ 20,668	$ 19,713	$ 19,738	$ 17,593	$ 23,615	$ 4,242	53
Pay the Rent	2010	Cash	16	10	6	63%	$100,000	$100,000	$ 18,000	$ 10,000	$ 5,000	$100,000	$ 11,250	2
Pick-A-Pair	1982	Price Guess	29	19	10	66%	$ 8,764	$ 8,558	$ 8,893	$ 7,800	$ 5,046	$ 16,675	$ 5,110	40
Pick-A-Number	1992	Price Guess	41	23	18	56%	$ 9,473	$ 8,175	$ 9,201	$ 8,160	$ 5,499	$ 17,523	$ 4,578	43
Plinko	1983	Cash	47	24	23	51%	$ 59,411	$ 50,194	$ 20,603	$ 12,190	$ 10,060	$202,165	$ 7,054	20
Pocket Change	2005	Car Pricing	19	7	12	37%	$ 22,754	$ 20,613	$ 19,805	$ 19,823	$ 16,837	$ 21,654	$ 7,303	19
Punch-A-Bunch	1978	Cash	29	7	22	24%	$ 25,810	$ 25,288	$ 5,944	$ 5,273	$ 5,142	$ 10,131	$ 2,378	73
Push Over	1999	Price Guess	49	36	13	73%	$ 8,780	$ 7,642	$ 9,213	$ 7,956	$ 5,438	$ 21,920	$ 5,845	32
Race Game	1974	4-Item	17	11	6	65%	$ 8,587	$ 8,586	$ 8,337	$ 8,444	$ 5,249	$ 11,568	$ 6,331	28
Range Game	1973	Price Guess	36	26	10	72%	$ 11,000	$ 10,230	$ 10,515	$ 9,765	$ 5,724	$ 23,250	$ 7,052	21
Rat Race	2010	Other Car	20	11	9	55%	$ 21,676	$ 21,791	$ 19,321	$ 19,169	$ 16,614	$ 23,346	$ 10,964	5
Safe Crackers	1976	Price Guess	18	12	6	67%	$ 9,846	$ 9,480	$ 10,320	$ 9,677	$ 5,675	$ 16,835	$ 6,451	27
Secret X	1977	Small Prize	30	16	14	53%	$ 10,501	$ 8,443	$ 11,563	$ 11,267	$ 6,930	$ 24,463	$ 6,034	29
Shopping Spree	1996	4-Item	15	8	7	53%	$ 8,319	$ 8,047	$ 8,592	$ 8,557	$ 7,412	$ 10,873	$ 4,563	44
Shell Game	1974	Small Prize	10	7	3	70%	$ 8,940	$ 8,502	$ 11,577	$ 10,422	$ 5,758	$ 21,417	$ 7,338	18
Side by Side	1994	Price Guess	52	31	21	60%	$ 7,578	$ 7,373	$ 7,750	$ 7,594	$ 5,076	$ 9,972	$ 4,527	47
Spelling Bee	1988	Other Car	19	9	10	47%	$ 21,989	$ 18,162	$ 17,955	$ 18,141	$ 15,251	$ 20,475	$ 9,407	9
Squeeze Play	1977	Price Guess	57	34	23	60%	$ 9,195	$ 8,295	$ 8,331	$ 7,626	$ 5,246	$ 19,258	$ 4,549	46
Stack the Deck	2006	Car Pricing	18	3	15	17%	$ 21,405	$ 20,668	$ 23,103	$ 24,183	$ 19,478	$ 25,649	$ 4,031	57
Swap Meet	1991	4-Item	15	9	6	60%	$ 7,366	$ 7,398	$ 7,739	$ 7,534	$ 6,397	$ 10,050	$ 4,520	48
Switch	1992	2-Item	45	28	17	62%	$ 9,268	$ 7,298	$ 10,087	$ 8,016	$ 5,219	$ 24,801	$ 4,988	41
Switcheroo	1976	Car Pricing	19	4	15	21%	$ 19,791	$ 20,129	$ 20,029	$ 20,718	$ 17,315	$ 21,365	$ 4,362	51
Take Two	1978	4-Item	15	8	7	53%	$ 8,157	$ 7,893	$ 7,800	$ 7,785	$ 6,998	$ 8,622	$ 4,152	54
Temptation	1973	Car Plus	15	5	10	33%	$ 27,527	$ 26,523	$ 6,571	$ 5,360	$ 4,940	$ 11,470	$ 2,666	71
Ten Chances	1975	Car Pricing	10	4	6	40%	$ 12,455	$ 19,645	$ 20,000	$ 20,070	$ 19,500	$ 20,360	$ 8,518	11
That's Too Much	2001	Car Pricing	49	11	38	22%	$ 24,910	$ 22,852	$ 25,605	$ 23,300	$ 19,291	$ 51,700	$ 5,231	38
Three Strikes*	1976	Car Plus	8	0	8	0%	$ 56,857	$ 57,993	$ -	$ -	$ -	$ -	N/A	N/A
Time is Money	2003	Cash	28	9	19	32%	$ 20,000	$ 20,000	$ 16,958	$ 18,914	$ 12,383	$ 20,000	$ 5,638	34
Triple Play*	2000	Car Plus	7	1	6	14%	$ 56,503	$ 57,694	$ 54,435	$ 54,435	$ 54,435	$ 54,435	N/A	N/A
2 for the Price of 1	1989	Price Guess	16	6	10	38%	$ 9,429	$ 9,073	$ 7,590	$ 7,341	$ 5,039	$ 10,540	$ 2,753	68
Vend-O-Price	2015	Grocery	33	14	19	42%	$ 9,786	$ 8,753	$ 9,705	$ 7,721	$ 5,438	$ 26,340	$ 3,275	63
Total/Wtd Avg			2131	1047	1084	49.1%	$ 16,181	$ 15,176	$ 12,322	$ 11,218			$ 5,718	

* Median Expected Win and associated Rank not shown for games played fewer than 10 times in seasons 47 and 48. $ Let's Make a Deal Mash Up games excluded.
Note: In the text of the book, I round individual pricing game statistics to the nearest $50 so as to not convey a false sense of precision over a limited data set.

Median Earnings
>$10,000
$8,000 - $10,000
$6,000 - $8,000
$4,000 - $6,000
$2,000 - $4,000

10 Highest Win Rate Games

Figure 10.2 Pricing Games—Key Statistics

games had the highest median earnings, driven by high win percentages in *Bonus Game*, *Cliffhanger*, and *Shell Game*. Cash games fell in the middle of non-car games in terms of median earnings, although contestants got to enjoy cold, hard cash. Three item pricing games had the lowest median earnings. Here the

Category	# Times Played	# Times Won	# Times Lost	Win Rate	Average Prize	Median Prize	Average Win	Median Win	Median Earnings	Earnings Rank
Car Pricing	483	172	311	36%	$ 21,517	$ 21,142	$ 21,363	$ 20,939	$ 7,509	2
Other Car	94	42	52	45%	$ 23,112	$ 19,917	$ 22,232	$ 18,515	$ 9,057	1
Car Plus	97	29	68	30%	$ 34,009	$ 34,265	$ 16,051	$ 14,104	$ 3,965	8
Total Car Games	674	243	431	36%	$ 23,537	$ 22,860	$ 20,879	$ 19,704	$ 7,215	
Cash	243	88	155	36%	$ 31,608	$ 28,508	$ 14,743	$ 11,422	$ 4,867	5
Price Guess	530	310	220	58%	$ 9,009	$ 8,419	$ 8,912	$ 8,319	$ 4,870	7
2-Item	226	150	76	66%	$ 8,368	$ 7,142	$ 8,706	$ 7,435	$ 4,935	6
3-Item	128	60	68	47%	$ 7,737	$ 7,533	$ 7,812	$ 7,516	$ 3,523	10
4-Item	80	41	39	51%	$ 8,268	$ 8,099	$ 8,290	$ 8,149	$ 4,361	9
Grocery	163	93	70	57%	$ 10,177	$ 9,412	$ 9,869	$ 9,003	$ 5,400	4
Small Prize	87	62	25	71%	$ 11,030	$ 9,010	$ 11,857	$ 10,259	$ 7,331	3
Total	2131	1047	1084	49%	$ 16,181	$ 15,176	$ 12,322	$ 11,218	$ 5,718	

Figure 10.3 Pricing Games—Key Statistics by Category

prize value was not large, the games were surprisingly difficult to win, and each game was all-or-nothing.

Different types of pricing games are more weighted toward different parts of the show. Sixty percent of car pricing games were played in the third, fifth, or sixth pricing game. In contrast, 43% of simple price guess games and 48% of two item pricing games—each usually played for a sub-$10,000 prize—were played in the second or fourth pricing games. Cash games were played more frequently in the first pricing game, perhaps to generate excitement at the start of the show. I suspect that car games are less likely in the first two pricing games so as to not diminish the chances of a contestant playing for a car if he is called into Contestant's Row later in the show.

There is some variation in prize values and win rate across the show, which is mostly tied to the variation in the frequency of car games across the show (Figure 10.4). Average prize value and average win value were highest in the first, third, and fifth games, and lowest in the second and fourth games, with some offset from a higher win rate in the fourth pricing game.

The first pricing game tended to have many big prizes to get the show started with gusto. Eight out of sixteen iterations of *Pay the Rent* were in the first pricing game and never in the second half of the show. With that said, average win value in the first pricing game would have averaged $14,000 without two *Let 'Em Roll* games played and won for $100,000. The fourth pricing game tended to have shorter run times, which ruled out many of the more time-consuming games. *Cover Up* and the *Card Game* were each played only once in the fourth pricing game.

Pricing games are the most lucrative part of the show. With median earnings value of more than $5,700, winnings in pricing games eclipsed winnings

Pricing Game	# of Wins	Win Rate	Average Prize	Average Win	Car Games	Cash Games	All Other Games
1	171	48%	$ 20,075	$ 15,210	111	58	187
2	175	49%	$ 14,467	$ 10,718	95	25	236
3	160	45%	$ 17,477	$ 12,804	133	57	166
4	192	54%	$ 12,183	$ 10,240	44	36	276
5	178	50%	$ 15,599	$ 13,665	147	20	189
6	173	49%	$ 17,302	$ 11,583	144	47	165
	1049	49%	$ 16,184	$ 12,323	674	243	1219

Figure 10.4 Pricing Game Performance across the Show

in Contestant's Row by a factor of four and in total eclipsed the $29,000 average Showcase winnings. More than $13 million was given away across 2,136 pricing games in seasons 47 and 48 for an average of $6,200 per pricing game and more than $37,000 per show.

It is rare for all six contestants to win their pricing games. It occurred only three times during seasons 47 and 48, and four times no pricing games were won. Figure 10.5 shows the distribution of pricing game wins per show and the expected frequency per the binomial distribution. Based on the observed 49% win rate, the binomial distribution models the probability of k wins across n pricing games (here 6 games) as $\binom{n}{k} p^k (1 - p)^{n-k}$ with probability p = 49%. The observed frequency and modeled frequency are quite similar, except for the more extreme outcomes.

Because each show usually includes a mix of hard-to-win games and easy-to-win games, six wins and six losses are less likely than predicted by

Wins per Show	Observed Count	Observed Frequency	Binomial Distribution	
			Predicted Count	Modeled Probability
0	4	1.1%	6	1.8%
1	38	10.7%	36	10.1%
2	82	23.0%	87	24.4%
3	118	33.1%	111	31.2%
4	80	22.5%	80	22.5%
5	31	8.7%	31	8.6%
6	3	0.8%	5	1.4%
	356	100%	356	100%

Figure 10.5 Pricing Game Wins per Show

the binomial distribution. After all, the 49% average win rate averages pricing games with win rates ranging from 20 to 80%. Put an 80% win rate game in the mix, and zero wins is very unlikely. Put a 20% win rate game in the mix, and six wins is very unlikely.

None of the aforementioned analysis suggests that a contestant should try and game their departure from Contestant's Row.[§] The difference in average prize value and win rate across pricing games is too modest to wait out the second or the fourth pricing game.

INTRODUCTION TO HEURISTICS AND PATTERN RECOGNITION

Heuristics are shorthand rules we humans follow that simplify our decision making when we lack the time, space, or resolve to independently analyze a situation or problem. Heuristics guided humanity when human life was primarily about survival, and it still serves us well in that capacity. It evolved in humans long before computers, factories, middle managers, and stock markets. Although there are many types of heuristics, rule-of-thumb heuristics, whereby one makes an approximation without doing exhaustive research, are most relevant for *The Price Is Right*.

Pattern recognition is about detecting patterns that stand a good chance of not having occurred at random, even if we do not understand the root cause. It is increasingly the realm of artificial intelligence, whereby computers use vast computing power, linear regression, and linear algebra to detect, test, and refine patterns that are then used for advanced computations and predictions—for example, in facial recognition.

The term heuristics generally comes with a negative connotation because the term focuses on the shortcuts that often lead us astray in our conclusions and decision making. Pattern recognition generally comes with a positive connotation, as it picks up on relationships and heuristics that are statistically significant and applies quantitative techniques to continuously test their validity. Statistically significant heuristics should do us more good than harm, provided we discard them when they are no longer predictive.

[§]Intentionally losing in the second bidding round would lead to a 1/4 chance of prevailing in the third bidding round, a 3/16 chance in the fourth round, a 9/64 chance in the fifth round, and a 27/256 chance in the sixth round. Multiplying those probabilities by the average win value in the third, fourth, fifth, and sixth pricing games would suggest an expected win value of $8,232 in later pricing games should the contestant intentionally fail to win the second bidding round.

TPIR Heuristics

Many pricing games on *TPIR* have strategies that allow for superior play, like targeting a total price guess in *Check Out*. Just as many have patterns that, if observed and followed, should materially improve contestant odds, like guessing only prize numbers whose last digit is zero in *Ten Chances*. *These heuristics work only if the showrunners continue to use them.* Some heuristics can be very powerful in pricing games, as detailed in chapters 11 through 21. My top ten are worth summarizing, along with quantification of the benefit they bestow on contestants.**

1. *Ten Chances:* Zero was the last digit of every two-digit, three-digit, and car prize, so choose only zero for the last digit. A contestant has virtually no chance of winning the car without using this heuristic, but a 90% chance of winning the car when doing so.
2. *Cliffhanger.* Guessing $25 for the first small item, $35 for the second, and $45 for the third would have always led to victory and with plenty of steps for the mountain climber to spare. *Cliffhanger* is easy without this heuristic but should be foolproof with this heuristic.
3. *One Away:* Digits two through five never repeated in the price of a car. Avoiding repeating numbers can improve odds of victory from 31% to as low as 44% or as high as 100%.
4. *Bonkers:* The correct choice for digits three through five was higher for displayed numbers one through five and lower for displayed numbers six through nine almost 100% of the time. Victory should be ensured in a few tries.
5. *Flip Flop:* Never was the right choice to flip and flop, reducing the number of choices from three to two for a four-digit prize.
6. *Squeeze Play:* Never was the right choice to squeeze the next-to-last number, reducing the number of choices from three to two for a four-digit prize.
7. *Money Game:* One choice is always the number of the current season of *TPIR*, but that choice was never correct. Avoiding this choice lifts odds from north of 50% to north of 60%.
8. *Safe Crackers:* The price of the three-digit prize always ended in zero.
9. *2 for the Price of 1:* The price of the three-digit prize always ended in zero when zero was an option, which was almost always the case. Guessing zero as the last number doubles the odds to well north of 60% in *Safe Crackers* and *2 for the Price of 1*.

**Although my heuristic observations are based on data from seasons 47 and 48, a quick review of results in season 49 based on data from third-party websites supports the continuity of these heuristics.

10. *That's Too Much:* The lowest two options and the highest two options were never the right pick. Moreover, *TPIR* seemed to have a preference for the third and seventh choices, which were correct 35% and 24% of the time respectively.

Rule of thumb heuristics are present in multiple avenues of life. *The Princeton Review* developed curriculums to improve performance on the SAT that in part focus on heuristics for eliminating incorrect options so that one can make an educated guess among a narrower set of choices. Doctors regularly use rule of thumb to diagnose illnesses, not having the time or leeway to run all the desired tests. I recall being misdiagnosed with an outer ear infection in my thirties because adults rarely develop middle ear infections and because there were signs of an external infection. In stock markets, rule-of-thumb heuristics abound. For example, the price-to-earnings multiple of the S&P 500 tends to vary predictably based on the underlying rate of inflation. An April 2021 CPI inflation reading surpassing 4%-plus year-on-year triggered a market sell-off, since 4%-plus inflation has historically led to market pullbacks. Curiously, this pullback proved temporary, even against a subsequent high inflation reading in May 2021. The key in effectively applying heuristics is to monitor their performance to gauge why and when they may be wrong. The May 2021 inflation reading occurred after the market's inflation expectations had settled down, causing it to be perceived as temporary rather than predictive. The doctor who misdiagnosed my ear infection said to call in three days if ear drops did not resolve my condition.

11

Two Item Pricing Games

Two item pricing games are among the most straightforward and "skill-less" pricing games on *The Price Is Right* (*TPIR*). The median prize values are lower, the probability of winning is higher, and the games are typically both short in duration and simple in their setup. Median win values ranged from a low of $6,250 in *One Right Price* to a high of $8,800 in *Do the Math*. Despite lower prize values, the high 66% win rates caused these games to fall only slightly below the middle of the pack in terms of median earnings among the seventy-four ranked pricing games (excluding *Golden Road*, *Three Strikes*, and *Triple Play*).

Category-wide median earnings of $4,950 and all other statistics in Figure 11.1 are rolled into in a category-weighted average at the bottom, which weights the underlying statistic from each game based on the number of times it was played. Average win value was $8,700 on account of a $32,000-plus win in *Bargain Game* during season 47's Dream Car Week, with two cars at stake, and a $25,000 win in *Switch* during season 47's Kids Week, with a trip for four to Costa Rica. The $7,450 median win value is more representative of performance.

Two item pricing games were played on 226 occasions during seasons 47 and 48, representing 10.6% of games played and occurring about three times in every five shows. Multiple two item pricing games were never played on the same show, which suggests that *TPIR* also sees these games as similar in nature.

One convention I employ in my discussion of two item pricing games and other pricing games is the percentage differential; for example, between a high price and a low price, as the difference relative to the average of the two prices. This approach is symmetric whether measuring the high number versus the low number or the low number versus the high number. For example, if a $5,000 trip with a displayed price of $7,000 in *Bargain Game* is a

101

Game	Played	Wins	Losses	Win Rate	Average Prize	Median Prize	Average Win	Median Win	Min Win	Max Win	Median Earnings	Earnings Rank
Bargain Game	51	41	10	80%	$ 8,624	$ 7,355	$ 8,744	$ 7,355	$ 5,278	$ 32,548	$ 5,913	31
Do the Math	48	31	17	65%	$ 9,129	$ 8,157	$ 9,850	$ 8,778	$ 6,464	$ 16,530	$ 5,669	33
Double Cross	23	14	9	61%	$ 7,090	$ 6,485	$ 7,387	$ 6,411	$ 5,548	$ 14,023	$ 3,902	59
Magic Number	13	11	2	85%	$ 6,778	$ 6,445	$ 6,913	$ 6,477	$ 5,448	$ 9,262	$ 5,481	35
One Right Price	46	25	21	54%	$ 7,497	$ 6,220	$ 7,206	$ 6,245	$ 5,216	$ 12,623	$ 3,394	62
Switch	45	28	17	62%	$ 9,268	$ 7,298	$ 10,087	$ 8,016	$ 5,219	$ 24,801	$ 4,988	41
Total/Wtd Avg	226	150	76	66%	$ 8,368	$ 7,142	$ 8,706	$ 7,435			$ 4,935	

Figure 11.1 Statistical Summary—Two Item Pricing Games

$2,000 bargain, then it is priced at a $2,000/($6,000) = 33% discount of the displayed price, and the displayed price is a 33% premium to the actual price.[*]

Two item pricing games are played as follows:

- *Bargain Game*: A contestant is shown two prizes, each with a "bargain" price that is below its actual retail price. He guesses which prize is the bigger bargain to the actual retail price. If he guesses correctly, he wins both prizes.
- *Do the Math*: A contestant is shown two prizes and the dollar value of their difference. He chooses whether the first prize plus the "difference" is equal in value to the second prize or whether the first prize minus the "difference" is equal in value to the second prize. In other words, which prize is more expensive. If correct, the contestant wins both prizes and the cash dollar value of the difference.
- *Double Cross*: Two strips of seven numbers dissect diagonally at the middle. The strip running from the upper left to the lower right contains the four-digit prize value of one of the prizes, and the strip running from the lower left to the upper right contains the four-digit prize value of the other prize. The contestant moves the highlighted number bands to select the four-digit prices he thinks correspond to the actual retail prices of the prizes. If he is right, he wins both prizes. There are four possibilities in each seven-number strip and in total, as the bands move simultaneously.
- *Magic Number*: The contestant is shown two prizes and is told which is the least expensive and which is the most expensive of the two. To win both prizes, he must move a digital number dial up or down to a "magic number," which must fall between the two prices.
- *One Right Price*: A contestant is shown two prizes and one price tag. He must choose which prize corresponds to the retail price shown on the price tag.
- *Switch*: A contestant is shown two prizes, each with an associated price tag. He must decide if the price tags are matched with the correct item or should be switched.

---

[*]The mathematically rigorous and symmetric way to calculate a percentage change is $\ln(y/x)$. The difference in two prices relative to the average of those prices provides a close approximation while offering ease of calculation.

BARGAIN GAME

Bargain Game stood out with its high 80% win rate (41 out of 51), ranking sixth across all pricing games, with contestants quite adept at identifying the bigger bargain. The win rate was far superior to the 54% in *One Right Price*, in which contestants struggled to evaluate absolute prices. Two trips are a popular setup, although prizes from recreational vehicles to big-screen televisions are also standard fare.

Twenty-eight times the item with the lower bargain price was the bigger bargain, and twenty-three times the item with the higher bargain price was the bigger bargain. The *difference* in the discounts ranged from $400 on the low end to $1,500 on the high end, with a median *discount difference* of $800. The two prizes tended to be priced closely together, with a median price difference of $900.

Smaller differences in discounts did not make mistakes more likely. Nor were mistakes more likely with lower or higher prize values. Mistakes tended to occur when items were priced outside the typical price range. On May 22, 2019, eight pairs of Tory Burch shoes with a bargain price of $1,704 were juxtaposed against a washer/dryer with a bargain price of $2,100. The contestant erred in picking the Tory Burch shoes, which were discounted by $600, versus the washer/dryer, discounted by $1,600. Needless to say, the $3,700 washer/dryer was an expensively priced set.

Two vacation prizes were offered in ten out of fifty-one iterations of the *Bargain Game*. In eight of these games, the farther destination was the bigger bargain. All airfares on *TPIR* are based on travel from Los Angeles, something not always internalized by contestants.

DO THE MATH AND SWITCH: BEWARE REVERSE PSYCHOLOGY

Do the Math and *Switch* each come down to identifying the more expensive prize. A 65% win rate (31 out of 48) in *Do the Math* suggests that most contestants were not confused by the computational setup. The 62% win rate (28 out of 45) in *Switch* was similar. One might have expected the win rate in *Switch* to be higher given a median price differential of $1,120 in *Switch* versus $885 in *Do the Math* and a median percentage delta of 30% versus 22%. Moreover, in *Switch* the contestant was provided with actual prices, not just the price difference.

In *Do the Math*, when the price differential was more than $700, contestants prevailed thirty out of forty-one times (73%). The higher price differentials allowed contestants to reach clearer conclusions as to which prize was more expensive. Conversely, when the price differential was less than $700, contestants struggled, winning only once in seven tries. Although the lower price differentials tended to occur with lower valued prizes, the setup still challenged contestants. On September 14, 2019, six Coach handbags were offered alongside a French-door refrigerator with a $614 price differential. The contestant incorrectly guessed that the refrigerator was less expensive than the handbags, which turned out to be priced at $3,499 and $2,885 respectively. An unwise choice, but a tough decision in light of the smaller price differential.

The lower win rate in *Switch* seemed to be driven by a disposition against switching. Contestants switched prices in only twenty out of forty-five games, yet the correct decision would have been to switch prices in thirty out of forty-five contests. It appears that some contestants employ reverse psychology and think they are being baited into mistakenly switching prices, or they mentally associate the displayed prices with the corresponding prizes.

> **Reverse psychology** is a well-studied form of behavior. Humans like to do the opposite of what they are told to do. Some companies use this to their advantage in marketing. Perhaps no example stands out more than the Patagonia holiday ad campaign in 2011, in which they told folks "do not buy this jacket" in a full-page Black Friday ad in the *New York Times*. Before we give Patagonia too much credit for their mastery of human psychology, note that this was part of their sustainability campaign. Still, with their jackets lasting close to three times that of competitive offerings, they could have their cake and eat it too—getting consumers to buy their jacket because they were told not to and driving a purchase decision with sustainability advantages.

Consider three losing examples in which it would have seemed straightforward to switch prices. On October 22, 2018, the contestant chose not to switch prices after seeing a grill with a displayed price of $2,500 and a fifty-five-inch 4K TV with a displayed price of $3,649. Switching would have meant an expensive $3,649 grill, but a realistically priced $2,500 fifty-five-inch 4K TV. On February 1, 2019, a contestant lost after not switching a $5,541 five-night trip to Miami with a $6,896 five-night trip to New Orleans, despite Miami being a farther flight and a more expensive destination. And on March 7, 2020, a contestant failed to switch Kors and Burberry accessories (six luxury items listed for $4,349) with a motorcycle listed at $3,434. When is a motorcycle priced below $3,500?

These losses seem harder to explain without assuming a propensity against switching on the part of contestants. Switching prices every time would have led to a 67% win rate, which is hard to attribute to chance. If there were 50/50 odds of a switch versus no-switch outcome, thirty or more switches or thirty or more "no switches" out of forty-five trials would occur only 3.6% of the time. It seems that the showrunners likely intended for there to be a disproportionate number of switching outcomes.

Key Tip

Focus on which prize is higher priced rather than on whether to switch or not to switch. Do not be averse to switching, given it was the right option two-thirds of the time during seasons 47 and 48.

DOUBLE CROSS

New in season 47, *Double Cross* challenged contestants. It was won fourteen times in twenty-three tries, with modest median earnings of $3,900. The total prize value fell between $5,500 and $8,000, except for two vacation prizes. This knowledge would have allowed contestants to eliminate two of four choices in nine of twenty-one non-vacation games and one of four choices in three of ten other games. Further eliminating unrealistic individual prices would have almost always reduced the game to two reasonable choices.

Only twice in twenty-three games was the correct pick to keep the original prices, and the correct choice was never the opposite ends of the price band. Contestants themselves favored the middle options, never choosing the original prices and only three times choosing the opposite ends. The contestant is better off choosing one of the middle options unless confident otherwise. Beware that the dual price bands can confuse contestants, who focus on the price of one item and not the other, leading to unforced errors.

Key Tip

Eliminate options with unrealistic individual prices from the four pricing pairings as well as pairings whose totals are too high or too low. Select the best remaining choice, favoring middle options.

MAGIC NUMBER

Magic Number was one of the easiest games, with eleven wins in thirteen tries but one of the lowest median prize values. Finding a number between two disparately priced items is not difficult. The higher priced prize had a $3,050 median premium to the lower priced prize, with that premium ranging from $2,200 to $6,100. Because the prices for lower and higher priced items tend to be quite clustered, *a "magic number" between $2,100 and $3,850 would have led to victories in each pricing game in seasons 47 and 48.* How's that for a heuristic!

The two losses featured poor play, with the player choosing a magic number that was too low. Notable here was the loss on October 3, 2018, when the contestant chose a magic number of $1,000 to fall between the price of a twenty-eight-inch personal computer and an ATV, which were priced at $1,700 and $4,644 respectively.

Even in winning, contestants chose a magic number close in price to the less expensive prize. In eleven wins, the magic number averaged $800 above the price of the less expensive prize and $2,900 below the price of the more expensive prize. In not one win was the magic number closer in price to the more expensive prize. Contestants would be well served to orient to the more expensive prize—or just choose $3,000 as their magic number.

> **Key Tip**
>
> First contemplate $3,000 as the magic number, and mentally test $3,000 versus the higher priced item. Alter only if there is good reason *not* to go with $3,000.

ONE RIGHT PRICE

One Right Price had a low $6,200 median prize value and was won in only twenty-five out of forty-six contests (54%), driving the lowest median prize value among all seventy-seven pricing games. The low win rate was surprising, given that a coin toss would have led to a 50% chance of choosing the right prize to match the displayed price.

The contestant was shown the higher price in thirty-nine out of forty-six games. If he had just selected the higher priced item, his win rate would have correspondingly increased.

The delta in price between the two prizes ranged from $630 on the low end to $2,400 on the high end, with a median of $1,135. However, losses were triggered by confusing prices for electronics or appliances rather than a large price delta between the two prizes. Contestants won only half the time a refrigerator was in play (4 out of 8) or a TV was in play (8 out of 15).

> **Key Tip**
>
> Unless confident that the displayed price pertains to the lower priced item, assume the right price is the higher price and choose the item you believe to be higher priced.

12

Three Item Pricing Games

Three item pricing games are similar to two item pricing games but involve pricing three prizes. Accordingly, there are more possible pricing scenarios and more ways to be wrong.

Three item pricing games had the worst outcomes among our ten defined categories; median earnings of $3,525 ranked dead last, well below the $4,950 median earnings for two item pricing games, as shown in Figure 12.1. Both categories of games had a similar median win value of roughly $7,500. However, the 47% win rate for three item pricing games was far below the 66% win rate for two item pricing games. Only in *Most Expensive* did contestants win more than 50% of the time. *Easy as 1-2-3* and *Make Your Move* ranked seventy-second and seventieth among seventy-four pricing games in median earnings. Fortunately for contestants, three item pricing games were played only 128 times in 2,136 pricing games during seasons 47 and 48, about once every three shows.

Game	Played	Wins	Losses	Win Rate	Average Prize	Median Prize	Avg Win	Median Win	Min Win	Max Win	Median Earnings	Earnings Rank
Easy as 123	19	7	12	37%	$ 7,444	$ 7,262	$ 7,145	$ 6,943	$ 6,334	$ 8,219	$ 2,558	72
Make Your Move	16	6	10	38%	$ 7,590	$ 7,289	$ 7,346	$ 7,113	$ 5,978	$ 8,749	$ 2,667	70
Most Expensive	44	25	19	57%	$ 7,907	$ 7,590	$ 8,079	$ 7,598	$ 6,630	$ 16,227	$ 4,317	52
One Wrong Price	49	22	27	45%	$ 7,745	$ 7,668	$ 7,847	$ 7,714	$ 6,087	$ 10,204	$ 3,463	61
Total/Wtd Avg	128	60	68	47%	$ 7,737	$ 7,533	$ 7,812	$ 7,516			$ 3,523	

Figure 12.1 Statistical Summary—Three Item Pricing Games

- *Easy as 1-2-3*: A contestant is given three blocks, one with 1 on it, a second with 2 on it, and a third with 3 on it. He has to place the 1 block on the least expensive price, the 2 block on the second-most-expensive prize, and the 3 block on the most expensive price. If all three blocks are placed correctly, the contestant wins all three prizes. He cannot win just one prize.

107

- *Make Your Move*: The contestant is shown a board containing a string of nine digits representing the prices of three prizes placed consecutively: one that's a two-digit number (i.e., $10–$99), one that's a three-digit number (i.e., $100–$999), and one that's a four-digit number (i.e., $1,000–$9,999). He must use two unique numbers in the string for the two-digit prize, three numbers for the three-digit prize, and four digits for the four-digit prize. The contestant wins all three prizes if and only if all three prices are correct.
- *Most Expensive*: A contestant picks the most expensive of three prizes. A correct answer wins all three prizes.
- *One Wrong Price*: A contestant is shown three prizes, each with an associated price tag. He must choose which prize has the incorrect price tag. The incorrect price tag may be higher or lower than the actual retail price. If correct, the contestant wins all three prizes.

EASY AS 1-2-3: PROCESS OF ELIMINATION

The name *Easy as 1-2-3* is a misnomer, because the 37% win rate is very low, and the median prize value of $7,250 hardly compensates for the low win rate. Most contestants listen to the audience's urgings in *Easy as 1-2-3*, which may include switching blocks after initial placement if the crowd shows disagreement. Nonetheless, the win rate remained low.

There are six possible rankings; namely, three possible choices for the most expensive prize and two conditional choices for the least expensive prize once the most expensive prize is chosen. The remaining block then corresponds to the middle prize. So total combinations are $3 \times 2 = 6$. Typically, the most expensive price falls in the $3,000 to $4,000 range, the middle prize falls in the $2,000 to $3,000 range, and the least expensive prize falls in the $1,000 to $2,000 range. The least expensive prize and middle prize were typically $700 to $1,100 apart, with a similar price differential for the middle prize and the most expensive prize.

Why is *Easy as 1-2-3* so difficult? For starters, figuring out the price order is quite difficult. Second, contestants do not understand the most logical way to tackle the game. They usually struggle with which is the most expensive prize among two possible choices and which is the least expensive prize among two possible choices. If the contestant is undecided with respect to both choices, he would have a one in three chance of winning. But he can do better.

Consider being undecided as to which of two items is the most expensive and which of two items is the least expensive. Put the "3" block on the item you are confident is not the least expensive. It may or may not be the most expensive. Then put the "1" block on the item you are confident is not the most expensive. It may or may not be the least expensive. Put the "2" block on the remaining item. If an item is clearly not the least expensive, then one can eliminate the two choices for which that item has the least expensive rank, as seen in Figure 12.2. If an item is clearly not the most expensive, then one can eliminate one further choice, leaving three possible outcomes, for no worse than 33.3% odds. This is consistent with the 37% win rate observed.

Block	1	2	3		1	2	3		1	2	3
	A	B	C	*A not least*	A	B	C	*make B mid-*		*Three Choices*	
	A	C	B	*expensive*	A	C	B	*prized prize*		*Eliminated*	
Prizes	B	A	C		B	A	C				
A-C	B	C	A		B	C	A		B	C	A
	C	A	B	*C not most*	C	A	B		C	A	B
	C	B	A	*expensive*	C	B	A		C	B	A

Figure 12.2 *Easy as 1-2-3*—Possible Combinations

Process of elimination is more common in the classroom than in our adult lives. But in medicine, there are many conditions, some serious, that are diagnosed by process of elimination. Consider a gastroenterologist diagnosing irritable bowel syndrome (IBS), characterized by abdominal pain, bloating, diarrhea, or constipation. Cleary the doctor wants to rule out colon cancer or a twisted bowel, far more serious conditions, among other temporary behaviors that may aggravate the intestinal track. But ultimately the diagnosis of IBS is through process of elimination. The same holds for fibromyalgia, a condition causing musculoskeletal pain accompanied by fatigue, sleep, memory, and mood issues. In other medical situations, process of elimination is used to rule out more observable conditions versus those that are less observable, such as an outer ear versus a middle ear infection. During the COVID-19 pandemic, the first condition to rule out when sick was, of course, COVID.

But odds should run closer to 50% if one puts the middle block on the prize that cannot be ruled out as least expensive or most expensive. Why? Because although three choices still remain, the fact that item B cannot be ruled out as least expensive or most expensive should make its placement in the middle about twice as likely as its placement at the extremes. On average, that means a 50% likelihood of item B being the middle prize versus a 25% probability of it being the least expensive or most expensive prize. And in some games one can rule out two prizes as not being the least expensive or the most expensive, reducing the game to two choices.

On May 6, 2019, the contestant playing *Easy as 1-2-3* ranked a range and cutlery as least expensive, six pairs of Coach designer shoes as most expensive, and a sixty-five-inch 4K HDTV as the middle prize. By no means was this an easy setup. But it could have been made easier. Start with the designer shoes from Coach. They are likely not the most expensive prize, as Coach shoes are not ultra-luxury, so give them the "1" block. Now consider which prize is not the least expensive. A range plus cutlery set has an uncertain price, but one can be pretty confident that a sixty-five-inch 4K QLED HDTV (a large high-tech TV) is not the least expensive prize. So place the "3" block (i.e., most expensive) on the HDTV. The actual prices were $3,107 for the range plus cutlery set, $2,200 for the TV, and $1,335 for the six pairs of Coach shoes. The heuristic did not work, but it got us to a good guessing point.

In another example that aired January 22, 2019, the prizes were an exercise bike, Coach accessories (three handbags and three pairs of shoes), and a sixty-five-inch 4K OLED Smart HDTV. Here the exercise bike is not the most expensive, so it gets block "1." The TV is not least expensive, so it gets block "3." The Coach products cannot be ruled out as least or most expensive, so they get block 2. The contestant made these same choices and won, with the bike priced at $1,299, the Coach accessories priced at $2,535, and the TV priced at $3,300.

Although I cannot ensure that a contestant would win 50% of the time with my heuristic, I tried it out and would have won 47% of the time (9 out of 19).*

Key Tip

Identify a prize that is not the least expensive; rank it most expensive (i.e., block 3). Identify a prize that is not most expensive; rank it least expensive (i.e., block 1). Put block 2 on the remaining prize.

MAKE YOUR MOVE

Make Your Move had a poor 38% win rate, similar to *Easy as 1-2-3*. Its $7,300 median prize value was low, and its $2,650 median earnings ranked seventieth out of seventy-four pricing games. Still, the nine-number sequence seems more daunting than it really is.

To tackle *Make Your Move*, first identify the six possible combinations for the different prizes. Those would be the options whereby the four-digit item begins on the first, third, fourth, or sixth number. If the four-digit item begins

---

*Please visit www.popculturemath.com to see the technique tested across seasons 47 and 48.

on the first or the sixth number, there are two possibilities for the three-digit prize and the two-digit prize; namely, the two-digit prize preceding the three-digit prize or vice versa. Note that no prize can start on the second digit or end on the eighth digit in the sequence, as that would leave unused numbers and unpriced prizes.

Consider the episode that aired May 30, 2019, captured by the illustration in Figure 12.3, and mapped out in Figure 12.4. Note that the sequence has two high-digit numbers (7–9), four mid-range digits (4–6), and three low-digit numbers (1–3). It is unlikely that the four-digit prize will start with a number less than five, since total prize value in a pricing game never fell below $5,000. This means the Colorado trip must be $8,642 or $6,142. Next, consider the lower priced prize, an adapter, specifically a 400-watt charger. It is very unlikely

Figure 12.3 *Make Your Move. Illustration courtesy of Omar Faruk*

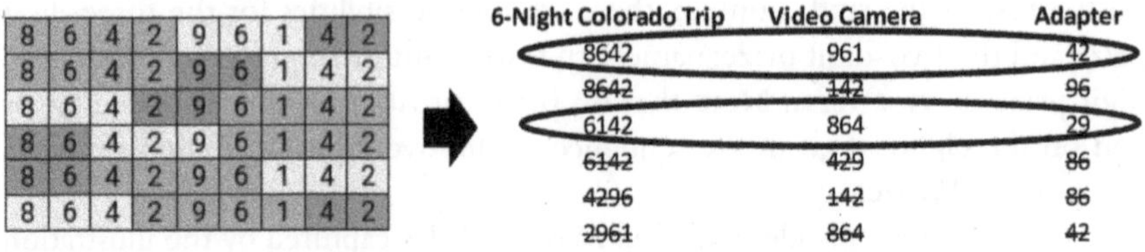

Figure 12.4 *Make Your Move* **Analysis—May 30, 2019, Episode**

that this prize will be more than $50, so we can rule out a $96 adapter and an $86 adapter. We are left with only two solutions that make sense: one with an $8,642 Colorado trip, a $961 video camera, and a $42 adapter, and one with a $6,142 Colorado trip, $864 video camera, and a $29 adapter. The contestant chose the option with the $6,142 Colorado trip and won.

Across the sixteen iterations of *Make Your Move*, nine had two choices that made sense and six had three choices that made sense. One had only one choice that made sense, allowing for an easy win. Assuming one in three odds with three reasonable choices, one in two odds with two reasonable choices, and a 100% chance of victory with one reasonable choice, one should have expected 7.5 wins in 16 iterations of *Make Your Move*, for a 47% win rate, versus six wins observed.

A contestant should be able to achieve a near 50% win rate in *Make Your Move* given some intuition as to which of the possible two- or four-digit choices are more likely. Only two of the ten losses involved the contestant choosing an unrealistic price combination. The other eight losses involved a realistic price choice but bad luck. Either way, *Make Your Move* should become more winnable with our strategy for eliminating unrealistic price options.

Key Tip

Identify the six possible combinations. Eliminate combinations that price the two-digit prize too low or too high or the four-digit prize too low or too high. Choose the best remaining option.

MOST EXPENSIVE

Choosing the most expensive prize is straightforward, and contestants were generally successful, with a 57% win rate (25 out of 44). Of the nineteen losses, eleven occurred when the contestant picked the mid-priced prize, and eight occurred when the contestant picked the least expensive prize. The most expensive prize and mid-priced prize differed by $600 to $1,500, and the mid-priced prize and the least expensive prize differed by $600 to $1,300.

The price differential between prizes did not meaningfully impact the win rate. Rather, contestants lost when they were confused about the price of one or more items. For example, on a show that aired June 19, 2019, a contestant chose a range plus cookware as the most expensive prize, passing over two electric bikes and a fifty-five-inch HD Smart TV. The unfamiliar electric bikes were the most expensive prize at $4,398, compared to $2,099 for the range plus cookware and $1,500 for the fifty-five-inch TV. In two other cases, contestants failed to correctly pick a home gym as the most expensive prize, also a difficult-to-price item. *Most Expensive* is a straightforward game but requires some pricing knowledge.

ONE WRONG PRICE: OUTLIER RECOGNITION

One Wrong Price is one of the most played games. Performance in the game was fair, with contestants winning 45% of the time (22 out of 49), whereas median earnings of $3,450 ranked an unimpressive sixty-first. In twenty games the wrong price was less than the actual retail price, and in twenty-nine games it was more than the actual retail price. Contestants struggled when the wrong price was less than the actual retail price, winning only seven out of twenty times, versus fifteen out of twenty-nine times when the wrong price was more than the actual retail price.

Unsurprisingly, the contestant was more likely to win when the wrong price was far off the actual retail price in percentage terms. When the price delta was less than 50% (measured relative to the average of the wrong price and the right price), contestants had five wins and fifteen losses. When the percentage delta was greater than 50%, contestants had seventeen wins and twelve losses.

One contestant pitfall was to shy away from choosing the most expensive prize as the wrong price. For example, on January 22, 2019, the contestant was shown a $1,999 Amigo scooter, a $3,999 1.3-gigahertz, twenty-eight-inch computer and printer, and a $4,599 Pro Form treadmill and digital audio player with headphones. All prices seemed a bit off, including the Amigo scooter (too low) and the computer/printer (too high). But when has a treadmill cost $4,599?

To improve performance in *One Wrong Price*, focus on the extreme ends of the price spectrum. All nineteen times that the low-priced item was priced wrong, it was priced too low. And fifteen of the nineteen times the high-priced item was priced wrong, it was priced too high. These setups would hardly ever occur by chance.

In fact, all twelve times the low-priced item was priced less than $1,000 (i.e., especially low), it was the wrong price. And twelve out of the sixteen times the higher priced item was priced more than $4,000 (i.e., especially high), it was the wrong price. This is very unlikely due to chance. A Monte Carlo simulation would lead to such outcomes (i.e., very low and very high prices being wrong twenty-four out of twenty-eight times) by chance only once in every thousand sets of trials. A contestant could have won twenty-four out of twenty-eight times by following this heuristic, in addition to a having a chance in the other twenty-one contests. Following this heuristic also would have made him less likely to have chosen the mid-priced item, which was the correct answer only 22% of the time. Remember, these heuristics are employed by the showrunners to make the game easier for contestants. Take advantage of them.

Key Tip

The lowest and highest priced items are more likely to be the wrong price. Specifically, look out for low-priced items priced too low and high-priced items priced too high, with sub-$1,000 items and $4,000-plus items very likely to be incorrectly priced.

Outlier Recognition

In the world of investing, outliers are often an indication that something is amiss, such as a company using aggressive financial reporting to inflate its results. During General Electric's tailspin at the end of former CEO Jeff Immelt's tenure, there was aggressive, likely illegal accounting. Most of this was through omission—namely, not disclosing the risk of losses in GE Capital's long-term care portfolio (i.e., insuring against chronic or terminal illness). But certain profit metrics stood out as outliers. Operating profit margins in the power business were near 20%, far above GE's competitors. And GE was using contract accounting to support those margins. GE reported profits related to servicing gas turbines and commercial aerospace engines based on the *average expected profitability* over the life of those long-term contracts. Assumptions for materially improving performance over time boosted that average. If that wasn't enough, GE routinely increased those estimates of expected profitability, boosting its earnings. These profits tied to contract assets were not supported by cash flow, but here GE Capital filled the cash flow gap at GE Industrial with cash infusions, including the factoring of long-term receivables. The long-term factoring with GE Capital was not reported despite being material. Since then, the Securities and Exchange Commission has cracked down on contract accounting, but for GE, the damage was already done.

13

Four Item Pricing Games

Four item pricing games were played infrequently, only a total of eighty times throughout seasons 47 and 48, or once every 4.5 episodes. These games have better win rates (51%) than three item pricing games (47%) but worse win rates than two item pricing games (66%). The four prizes typically total $7,000 to $10,000, with the median prize value of $8,100 exceeding that of three item pricing games ($7,550) and two item pricing games ($7,150).

Four item pricing games, notably *Race Game* and *Danger Price*, make for some of the more dynamic pricing games. Most four item pricing games fall in the middle of the pack in terms of median earnings, as seen in Figure 13.1, with *Race Game* having a better win rate (65%) and *Danger Price* a worse win rate (28%).

Game	Played	Wins	Losses	Win Rate	Average Prize	Median Prize	Avg Win	Median Win	Min Win	Max Win	Median Earnings	Earnings Rank
Danger Price	18	5	13	28%	$ 8,767	$ 8,436	$ 9,477	$ 8,539	$ 8,092	$ 12,656	$ 2,372	74
Race Game	17	11	6	65%	$ 8,587	$ 8,586	$ 8,337	$ 8,444	$ 5,249	$ 11,568	$ 6,331	28
Shopping Spree	15	8	7	53%	$ 8,319	$ 8,047	$ 8,592	$ 8,557	$ 7,412	$ 10,873	$ 4,563	44
Swap Meet	15	9	6	60%	$ 7,366	$ 7,398	$ 7,739	$ 7,534	$ 6,397	$ 10,050	$ 4,520	48
Take Two*	15	8	7	53%	$ 8,157	$ 7,893	$ 7,800	$ 7,785	$ 6,998	$ 8,622	$ 4,152	54
Total/Wtd Avg	80	41	39	51%	$ 8,268	$ 8,099	$ 8,290	$ 8,149			$ 4,361	

* Take Two often fails to reveal the prices of non-chosen prizes. In these cases, we estimated prize values based on comparable prizes.

Figure 13.1 Statistical Summary—Four Item Pricing Games

- *Danger Price*: A contestant is shown a "danger price," which is the actual price of one of the four prizes. He then picks the three items among the four prizes, one at a time, that do not correspond to the "danger price." If he avoids picking the item with the "danger price," he wins all four prizes.
- *Race Game*: The contestant is shown four prizes and given four price tags to match to the four items. The clock is set for forty-five seconds.

The contestant places the four prices on the four prizes and then pulls a lever on the clock to see how many prices—but not which ones—he has right. The clock continues to tick down while he attempts to correct his pricing errors by switching price tags and runs back to again pull the lever. The game ends when the contestant wins all four prizes or the forty-five seconds expire, in which case he wins the prizes he has correctly priced.

- *Shopping Spree*: The contestant is shown an amount of money he must exceed by picking the three most expensive among the four prizes, one at a time. If he picks all three without selecting the lowest priced prize, he surpasses the *Shopping Spree* limit and wins all four prizes.
- *Swap Meet*: The player is shown one prize ("swap item") but not its price, along with three other prizes. He has one chance to pick which of the other three prizes has the same value as the swap item.
- *Take Two*: The contestant is shown four prizes. He must pick the two items whose total matches a target price in order to win all four prizes. If incorrect, he gets one more try to win, now knowing two prices.

DANGER PRICE

Pray you do not play *Danger Price*. Contestants won only five out of eighteen times (28%) and median earnings of $2,350 ranked seventy-fourth, dead last. Usually there are one or two prizes that are clearly less expensive or more expensive than the danger price, but that still leaves two or three prizes that might be the danger price. So no real advice in *Danger Price*, except to do your best.

On the June 6, 2019, episode, the contestant avoided the danger price of $2,098 with his first pick, the $849 foosball table. However, he then picked the $2,098 washer/dryer instead of the Kate Spade accessories (one scarf, two bags, two pairs of shoes) priced at $1,090 or a sixty-five-inch 4K OLED HDTV priced at $3,300. Like the contestant, I too would have struggled after eliminating the foosball table.

The danger price was the most expensive item four times out of eighteen times, the second most expensive item nine times, the third most expensive item once, and the least expensive item four times. Having a danger price equal to the second or third most expensive prize makes the game even harder, as it is harder to eliminate options. The contestant won four out of eight times when the danger price corresponded to the most expensive or least expensive item, but only one out of ten times when the danger price was the second or third most expensive item.

RACE GAME: SCENARIO SAMPLING

Race Game is one of my favorite games on *The Price Is Right* (*TPIR*), even if the individual prizes tend to be a bit pedestrian. It combines an entertaining mix of skill, luck, and speed. There were eleven wins in seventeen games, placing *Race Game* twenty-eighth in median earnings. Two of those eleven wins had the contestant getting the two most expensive prizes right, surpassing $5,000. Of the six "losses," contestants twice won two prizes valued at less than $5,000, twice won one prize, and twice went home empty-handed.

Speedy contestants have time for four tries before the forty-five seconds expire, although slower contestants may have time for only three tries. Winning may seem like a daunting task, but strategic play can make the game considerably easier. There are twenty-four possible price combinations: four for the first prize, then three for the second prize, then two for the third prize, with the fourth prize then a given. Of the twenty-four pricing combinations, one combination wins all four prizes, six combinations win two prizes, eight combinations win one prize, and nine combinations win no prizes.

Consider the hypothetical example in Figure 13.2, in which Prize A is priced at $1,000, Prize B at $2,000, Price C at $3,000, and Prize D at $4,000. If the contestant knows that Prize A is not the most expensive and Prize D is not the least expensive, he can reduce the number of choices to fourteen by removing the ten combinations in which A is the most expensive and/or D is

24 Combinations

Prize Value				Correct	
$1,000	$2,000	$3,000	$4,000	# Prizes	
A	B	C	D	4	
A	B	D	C	2	
A	C	B	D	2	
A	C	D	B	1	*A not most*
A	D	B	C	1	*expensive*
A	D	C	B	2	
B	A	C	D	2	*D not least*
B	A	D	C	0	*expensive*
B	C	A	D	1	
B	C	D	A	0	
B	D	A	C	0	
B	D	C	A	1	
C	A	B	D	1	
C	A	D	B	0	
C	B	A	D	2	
C	B	D	A	1	
C	D	A	B	0	
C	D	B	A	0	
D	A	B	C	0	
D	A	C	B	1	
D	B	A	C	1	
D	B	C	A	2	
D	C	A	B	0	
D	C	B	A	0	

14 Combinations

Prize Value				Correct	
$1,000	$2,000	$3,000	$4,000	# Prizes	
A	B	C	D	4	
A	B	D	C	2	
A	C	B	D	2	*B not most*
A	C	D	B	1	*expensive*
A	D	B	C	1	
A	D	C	B	2	
B	A	C	D	2	*C not least*
B	A	D	C	0	*expensive*
B	C	A	D	1	
C	A	B	D	1	
C	A	D	B	0	
C	B	A	D	2	
C	D	A	B	0	

8 Combinations

Prize Value				Correct
$1,000	$2,000	$3,000	$4,000	# Prizes
A	B	C	D	4
A	B	D	C	2
A	C	B	D	2
A	D	B	C	1
B	A	C	D	2
B	A	D	C	0
B	C	A	D	1
B	D	A	C	0

A least expensive or D most expensive

6 Combinations

Prize Value				Correct
$1,000	$2,000	$3,000	$4,000	# Prizes
A	B	C	D	4
A	B	D	C	2
A	C	B	D	2
A	D	B	C	1
B	A	C	D	2
B	C	A	D	1

Figure 13.2 *Race Game* Combinations

the least expensive. Of these fourteen combinations, four result in zero prizes, four result in one prize, five result in two prizes, and one results in all four prizes. The contestant is much better off already.

If the contestant can further conclude that Prize B is not the most expensive and Prize C not the least expensive, he can further whittle the number of choices down to eight. Of these eight combinations, two result in zero prizes, two result in one prize, three result in two prizes, and one wins all four prizes. Finally, if he can further conclude that *either* Prize A is the least expensive or Prize D is the most expensive, his number of choices is reduced to six. Two of these choices result in one prize, three result in two prizes, and one wins all four prizes.

I believe contestants effectively face the eight scenario or six scenario *Race Game*, depending on the prizes and corresponding prices. On June 13, 2019, the contestant played for two paddleboards, a three-gigahertz personal computer, five Coach handbags, and a motorcycle, with price tags of $860, $1,898, $2,460, and $3,499, respectively. It seemed clear that the motorcycle and handbags were not $860, and the paddleboards and computer were not $3,499. Moreover, the motorcycle was likely $3,499. So only six likely pricing combinations remained. The contestant tried two of these six combinations and won on his second try.

Now reconsider the hypothetical example with Prizes A, B, C, and D priced $1,000, $2,000, $3,000, and $4,000 without knowing which price goes with which prize a priori. One can calculate the expected win value on the first try in the twenty-four combination game (no elimination), the fourteen combination game (modest elimination), the eight combination game (moderate elimination), and the six combination game (high elimination). One can also calculate expected earnings with two tries, three tries, and four tries, assuming the player does not stop after winning one or two prizes.

Figure 13.3 details the results and shows why contestants do well in *Race Game*. They can usually rule out two prizes being the most expensive and two prizes being the least expensive.[*] The result is eight pricing scenarios, and an expectation to win 55% of the total prize value (i.e., $5,536 out of $10,000) with up to three tries in forty-five seconds. If a least expensive or most expensive price is also clear, the expected win value increases to 70% of total prize value ($7,000 out of $10,000) with three tries.

Distribution of Results	Scenarios	0	1	2	4	1 Try	2 Tries	3 Tries	4 Tries
No elimination	24	9	8	6	1	$2,500	$2,826	$3,152	$3,478
A not most expensive, D not least expensive	14	4	4	5	1	$3,214	$3,736	$4,258	$4,780
A/B not most expensive, C/D not least expensive	8	2	2	3	1	$3,750	$4,643	$5,536	$6,429
And A least expensive or D most expensive	6	0	2	3	1	$5,000	$6,000	$7,000	$8,000

Column groups: **Correct Items - Scenario Count** (0, 1, 2, 4); **Expected Win Value** (1 Try, 2 Tries, 3 Tries, 4 Tries).

Figure 13.3 *Race Game* Expected Win Value

Median earnings in *Race Game* were $6,331 versus average prize value of $8,587, for a 74% conversion. This is most consistent with the six-scenario setup with three tries (i.e., winning 70% of total prize value) and suggests that contestants were able to identify a least expensive or most expensive prize in addition to ruling out two prizes as least expensive or most expensive.

When contestants won all four prizes, they won twice on their first guess, five times on their second guess, and twice on their third guess. The average try took fourteen seconds. The slowest contestant made only two tries in forty-five seconds, whereas the two fastest contestants executed two tries in twenty seconds, achieving a full win. Seven contestants erred in placing the high price on an item clearly not the most expensive or the lowest price on an item clearly not the least expensive. Two of those seven contestants ended up empty-handed, but two subsequently got on track and won all four prizes. The others won one or two prizes.

Key Tip

Before the clock starts, try to identify two prizes that are not the least expensive and two prizes that are not the most expensive. See if you can pick a least expensive or most expensive prize. Try feasible combinations, starting with those that seem more likely.

Scenario Sampling

In our busy lives, we are well-served to do quick scenario sampling to find out what works and what doesn't. But this is not so easily done. Speed dating tries to introduce individuals to many potential romantic partners through five-minute exchanges, but five minutes is hardly enough time to judge one's interests and attractions. Stitch Fix has figured out a formula for quick scenario sampling of ladies' (and men's) clothing. Subscribers receive a selection of five clothing items, typically every month, picked out by stylists and algorithms created by data scientists. Those women decide what they want to keep and what they want to return, with incentives to keep at least one item or all five items. Like *Race Game*, Stitch Fix narrows the number of reasonable options, which the customer then quickly samples.

SHOPPING SPREE

In *Shopping Spree*, the contestant must identify the least expensive prize and avoid picking that prize to reach the desired sum. This is an easier setup than *Danger Price*, in which the price to be avoided could have any price rank.

Contestants won *Shopping Spree* 53% of the time (8 out of 15), about twice as frequently as *Danger Price*. Moreover, there are usually only two realistic contenders for the least expensive prize, setting up odds that should be no worse than 50/50.

Like *Race Game*, the most expensive prize was more than $3,000 (median $3,500) and the least expensive prize was less than $1,000 (median $700), with one $1,350 exception. The second and third highest priced prizes fell in the $2,000 to $2,750 range (median $2,500), with one $3,100 exception, and the $1,300 to $1,850 range (median $1,630), with one $2,200-plus exception. To lose *Shopping Spree*, the contestant effectively would have to confuse an item priced at approximately $1,630 with an item priced at approximately $700. This occurred on May 20, 2019, when the player erred in picking the $798 cardio strider third instead of the $1,699 gas range. His *Shopping Spree* total of $6,947 fell short of the $7,500 target.

SWAP MEET

Swap Meet is a four item pricing game that functions like a three item pricing game, as only three prices are unknown. Contestants won *Swap Meet* nine out of fifteen times. The 60% win rate suggests that the contestant is able to rule out one unrealistic swap prize and then guess from the remaining two selections with better than 50/50 odds.

On June 11, 2019, the contestant had to pick out a prize with the same price as three pairs of Christian Louboutin shoes, which turned out to be priced at $2,435. The contestant picked the computer and printer, priced at $1,199, and lost. The correct pick was the Ping-Pong table, also priced at $2,435. The small kitchen appliances, priced at $505, were far too low in price and not a realistic option.

In *Swap Meet*, there are no useful heuristics. The swap prize corresponded to the low-priced prize in seven games, the mid-priced prize in five games, and the high-priced prize in the remaining three games. When the contestant lost, it was usually because an item had an abnormally low or high price.

TAKE TWO

In *Take Two*, the contestant must pick two out of four items, the sum prices of which equal a target price. *Take Two* was won 53% of the time (8 out of 15), with median earnings of $4,150, ranking fifty-fourth. There are six possible nonduplicative prize pairings. Item 1 can be paired with items 2–4, item 2 can be paired with items 3–4, and item 3 can be paired with item 4. The game is made easier by the fact that the contestant gets to see two prices after an initial incorrect guess.

Similar to other four item pricing games, *Take Two* tends to have prizes distributed across the sub-$1,000, $1,000 to $2,000, $2,000 to $3,000, and $3,000-plus ranges. In all but one iteration of *Take Two*, the first and third ranked prizes totaled the target price. Identifying which prizes fall into those price buckets was a more difficult setup.

In the episode that aired June 20, 2019, the target price of $5,160 required a prize priced at more than $3,000 to be paired with a prize in the $1,000 to $2,000 range. The contestant won on his first try. Had he not won and had both prices likely been in the wrong pricing category (i.e., neither first nor third), winning on the second try would have been automatic. If one of the prizes was in the right price category (i.e., either first or third), he would have had at least a 50/50 chance of pairing it with one of the two remaining prizes.

Consider the odds of success if the prize values stay in their typical ranges and the contestant knows which ranked prizes total the target but does not know how the prizes rank a priori. If his first pick is a random one-in-six guess, there is a one-in-five chance of a certain win on the second try, because one in five times neither prize selected first was in the right pricing category. The other four out of five times, the odds of winning should be no less than 50%, as the contestant keeps the prize in the right pricing category and chooses one of the other two prizes to sum to the desired total. This means at least a $\frac{3}{5}$ chance on the second try: $\frac{1}{5} + \left(\frac{1}{2} \times \frac{4}{5}\right) = \frac{3}{5}$. The probability of winning P(win) $> \frac{1}{6} + \left(\frac{5}{6} \times \frac{3}{5}\right) = \frac{2}{3}$.

Two games did not conform to this analysis: (1) one game in which the highest and lowest ranked prizes totaled the target price, and (2) one game in which the prizes were more expensive than usual, ($4,618, $3,799, $2,180, $1,210). The contestant lost in both cases.

Positively, contestants often have insights into the likely prices of the various prizes, increasing odds of winning relative to the aforementioned calculation. Four wins on the first pairing reflected better success than one in six random odds. However, four wins on eleven second tries was less impressive, suggesting that many contestants failed to consider which price buckets their initial picks fell into. On December 26, 2018, a contestant initially picked a $2,699 treadmill and $1,475 speakers, likely the second and third ranked prizes, falling short of the $4,974 target. On his second turn he tried the other two prizes, an $860 computer and a $3,499 refrigerator, also falling short of the target. Instead, he should have kept the $1,475 speaker (the third ranked prize) and picked the $3,499 refrigerator, the most expensive prize.

The showrunners likely set the target price equal to the sum of the first and third ranked prizes to make the game more difficult. Usually, the least expensive prize is the easiest to identify. Still, a target price equal to the sum of the highest and lowest priced items or the two middle-priced items should not be ruled out prospectively. Were this to happen, a lower target price should be the giveaway.

Key Tip

Is the target price realistically the sum of the first and third ranked prizes based on their typical price ranges? If yes, identify the likely rank of the initially selected prizes. Keep the likely first or third ranked prize, and pair it with the one of the other two prizes that is complementary. If no, switch both prizes.

14

Price Guess Games

Price guess games are the most common pricing games. They were played 530 times during seasons 47 and 48, comprising one out of four pricing games. Price guess games typically were played once or twice per show, although they could appear three times per show or not at all. *Double Prices*, *Flip Flop*, *Side by Side*, and *Squeeze Play* were four of the seven most frequently played pricing games on *The Price Is Right* (*TPIR*), each played more than fifty times. Price guess games were overrepresented in the second and fourth pricing games, appearing 115 times in each, and underrepresented in the third game, appearing only 47 times.

There was a wide range of win probabilities and median prize values across the fourteen price guess games, as shown in Figure 14.1. In aggregate,

Game	Played	Wins	Losses	Win Rate	Average Prize	Median Prize	Average Win	Median Win	Min Win	Max Win	Median Earnings	Earnings Rank
Two Price Choice												
Coming or Going	39	26	13	67%	$ 7,406	$ 7,619	$ 7,564	$ 7,678	$ 6,078	$ 9,056	$ 5,118	39
Double Prices	53	33	20	62%	$ 9,812	$ 9,498	$ 9,614	$ 9,619	$ 5,338	$18,298	$ 5,989	30
Flip Flop	52	31	21	60%	$ 7,780	$ 7,588	$ 7,606	$ 7,565	$ 5,719	$ 9,783	$ 4,510	49
Side by Side	52	31	21	60%	$ 7,578	$ 7,373	$ 7,750	$ 7,594	$ 5,076	$ 9,972	$ 4,527	47
Squeeze Play	57	34	23	60%	$ 9,195	$ 8,295	$ 8,331	$ 7,626	$ 5,246	$19,258	$ 4,549	46
Three or More Price Choice												
Balance Game	36	15	21	42%	$11,193	$10,304	$11,093	$ 9,938	$ 6,318	$18,215	$ 4,141	55
Freeze Frame	35	14	21	40%	$ 7,509	$ 7,429	$ 7,150	$ 6,790	$ 5,598	$ 9,148	$ 2,716	69
Pick-A-Number	41	23	18	56%	$ 9,473	$ 8,175	$ 9,201	$ 8,160	$ 5,499	$17,523	$ 4,578	43
Push Over	49	36	13	73%	$ 8,780	$ 7,642	$ 9,213	$ 7,956	$ 5,438	$21,920	$ 5,845	32
Secondary Prize Pricing												
Clock Game	13	8	5	62%	$13,804	$10,420	$15,184	$10,515	$ 7,369	$51,285	$ 6,668	24
Safe Crackers	18	12	6	67%	$ 9,846	$ 9,480	$10,320	$ 9,677	$ 5,675	$16,835	$ 6,451	27
2 for the Price of 1	16	6	10	38%	$ 9,429	$ 9,073	$ 7,590	$ 7,341	$ 5,039	$10,540	$ 2,753	68
Close in Price												
Bonkers	17	9	8	53%	$ 7,606	$ 7,593	$ 7,433	$ 7,549	$ 5,942	$ 9,349	$ 3,997	58
Check Game	16	6	10	38%	$ 7,875	$ 7,500	$ 8,094	$ 8,034	$ 7,550	$ 8,930	$ 3,013	66
Range Game	36	26	10	72%	$11,000	$10,230	$10,515	$ 9,765	$ 5,724	$23,250	$ 7,052	21
Total/Wtd Avg	530	310	220	58%	$ 9,009	$ 8,419	$ 8,912	$ 8,319			$ 4,870	

Figure 14.1 Statistical Summary—Price Guess Games

123

the 58% win probability compared to 49% for *TPIR* as a whole, and the median prize value of $8,400 was in line with pricing games not played for a car or for cash. Certain games like *Coming or Going* and *Push Over* resulted in high win rates, as they presented a choice between two pricing options, with one more likely than the other. Other games like *Freeze Frame* and *Balance Game* proved more difficult, because there were usually three, not two, realistic pricing choices.

I subdivided the fifteen price guess games into four subcategories: two price-choice games, three or more price-choice games, secondary prize pricing games, and games requiring one to be close in price.

TWO PRICE-CHOICE GAMES

The contestant chooses from two price choices.

- *Coming or Going*: A contestant is shown a four- or five-digit number that either equals the price of the prize or equals the price of the prize when the digits are sequenced in reverse order. Visually, the numbers come toward you or go away from you. If the contestant chooses the right price, he wins.
- *Double Prices*: The contestant chooses the price of the prize in question from two possible prices.
- *Flip Flop*: A contestant is shown a four-digit price for a prize which is the correct price when the first two numbers are reversed (i.e., flipped), the latter two numbers are reversed (i.e., flopped), or both the first two and last two numbers are reversed (i.e., flip flopped). The contestant decides whether to flip, flop, or flip and flop and wins if correct.
- *Side by Side*: Two pairs of two digits are stacked on top of each other. To win, the contestant must decide whether the top two digits should be moved to the right or left of the lower two digits to match the prize value.
- *Squeeze Play*: The contestant is given a set of five numbers (or occasionally six numbers). One of the middle numbers (the second through fourth or second through fifth) doesn't belong in the actual retail price and must be removed. Once a number is taken out, the remaining numbers squeeze together. If the price is correct, the contestant wins.

THREE OR MORE PRICE-CHOICE GAMES

The contestant chooses from three or more price choices.

- *Balance Game*: A contestant chooses two of three bags in $1,000 increments, which, when added to a sub-$1,000 residual dollar amount (also in a bag), equals the price of the game's prize. The bags are put on a balancing scale to see if the total price "balances" with the prize value.
- *Freeze Frame*: Eight two-digit number tiles rotate. A frame encloses two tiles at the top of the ring, which form a four-digit price. The player must freeze the frame at the correct combination to win.
- *Pick a Number*: A contestant is shown the price of a prize (sometimes a four-digit price and sometimes a five-digit price) with either the first or second digit missing. The contestant must choose the correct missing digit from among three possible numbers to win the game and the prize.
- *Push Over*: The contestant is shown a row of nine numbered blocks, which are comprised of the four- or five-digit prize value and other blocks that precede and follow those blocks. The last four or five digits are initially enclosed by a colored window. The contestant irretrievably pushes the blocks over one by one until his selected four- or five-digit price is in the window. If correct, he wins.

SECONDARY PRIZE PRICING GAMES

The contestant prices a three-digit prize to win a larger prize.

- *Clock Game*: To win the main prize, the contestant has thirty seconds to guess the price of two small prizes, each generally priced less than $1,000. He names a price and is told "higher" or "lower" by Drew, allowing him to adjust his guess and home in on the price to the exact dollar. If the contestant correctly guesses the prices for both prizes within thirty seconds, he wins both the smaller prizes and the main prize. If he gets one price correct, he wins the first small prize but not the main prize.

- *Safe Crackers*: The contestant must guess the price of a three-digit prize to win that prize and the larger prize, ranging from $5,000 to $15,000. Both prizes are locked in a safe. The contestant has three safe dials, each with the same three numbers. He must select the price of the three-digit prize using each number only once. He wins if the three-digit selection is correct, upon which the safe opens.
- *2 for the Price of 1*: To win the game's main prize, the contestant must correctly guess the price of a smaller three-digit prize. He is shown two possible numbers for the first digit, two numbers for the second digit, and two numbers for the third digit. He gets one digit of his choice revealed and must pick the other two digits from the two number choices. If correct, he wins the main prize.

CLOSE IN PRICE

The contestant must be sufficiently close in price or directionally correct to win.

- *Bonkers*: A contestant is shown a four-digit number, with certain digits in the price of the prize more than the displayed number and certain digits less than the displayed number. The contestant uses disks to designate which numbers in the price of the four-digit prize are more than the displayed numbers and which are less than the displayed numbers. He has thirty seconds across multiple tries to get the disks placed correctly before the clock runs out.
- *Check Game*: A contestant writes a check for an amount that, when added to the prize value, falls between $8,000 and $9,000. If correct, he wins the prize and the check's cash value.
- *Range Game*: A range finder spanning $150 moves upward electronically in a $600 range. The contestant must stop the range finder so that the price of the prize falls within the $150 range of the range finder.

COMING OR GOING, DOUBLE PRICES, AND SIDE BY SIDE: ANCHORING HIGH

In *Coming or Going*, *Double Prices*, and *Side by Side*, one chooses the price of the prize from two possible prices. In *Coming or Going*, the numbers swing to the right ("going") or come in reverse at the left ("coming"); for example, a choice between $8,966 and $6,698. In *Double Prices*, there are two price choices, one above the other. And in *Side by Side*, two digits are stacked on top of two other digits. The contestant must decide if the numbers on the top are the first two digits in the prize, moving them down and to the left, or the last two digits, moving them down and to the right. One contest for a trip to Bali had the numbers "62" above the numbers "85," with the contestant choosing between $6,285 and $8,562.

Contestants won 67% of the time in *Coming or Going* (26 out of 39 iterations), 62% of the time in *Double Prices* (33 out of 53), and 60% of the time in *Side by Side* (31 out of 52). Median prize values were $7,600 for *Coming or Going*, $9,500 for *Double Prices*, and $7,350 for *Side by Side*, leading to median earnings of $5,100, $6,000 and $4,550, respectively.

The difference in price between the two price options was similar across games. In *Coming and Going*, there was a 27% median delta between the two price options measured relative to the average of the two prices, compared to 30% in *Double Prices* and 27% in *Side by Side*. Interestingly, the contestant win rate was similar in each of the games regardless of whether the percentage price differential was low or high. The absolute dollar difference was higher in *Double Prices* (median $2,638) than in *Side by Side* (median $2,080) or *Coming or Going* (median $1,950), mainly due to a higher prize value.

What stood out in terms of contestant behavior in these three games was a tendency to choose the higher priced option. The contestant chose the higher priced option twenty-seven of thirty-nine times (69%) in *Coming or Going*, forty-three out of fifty-three times in *Double Prices* (81%), and thirty-one out of fifty-two times (60%) in *Side by Side*. The higher price option was always presented on top in *Double Prices*. In *Side by Side*, the higher digits were on top only 40 percent of the time, which may explain why the contestant chose the higher priced option less in *Side by Side* than in the other two games. Placement of the higher digits on top seemed to encourage the contestant to favor the higher priced option.

The right price was the higher price twenty-four out of thirty-nine times (62%) in *Coming or Going*, twenty-six out of fifty-three times in *Double Prices* (49%), and twenty-two out of fifty-two times (42%) in *Side by Side*. Hardly suggestive of any bias on the part of the showrunners.

What exactly explains the dichotomy between underbidding in Contestant's Row and the tendency to err high in *Coming or Going*, *Double Prices*, and *Side by Side*? I believe reference prices eliminate contestants' inherent underbidding biases by anchoring them around the price of the prize. Moreover, contestants are not likely to see *The Price Is Right* as cheap, which pushes them toward the higher priced option. The tendency to choose high was probably more pronounced in *Double Prices* because the high price was always on top.

In contrast, when there is a risk of going over and no price to anchor to, contestants tend to underestimate actual retail prices.

Key Tip

Do not assume a *TPIR* bias toward the more expensive price option in *Coming or Going, Double Prices,* or *Side by Side.*

Anchoring High

Sellers of goods, real estate, and businesses would love to anchor the buyer high. In the realm of luxury goods, sellers do just that. The practice, often called prestige pricing, relies on the premise that for certain luxury goods, a high price conveys quality and prestige. In the watch market, brands like Rolex, Omega, and Patek Philippe routinely sell watches for tens of thousands of dollars, whereas brands like Seiko, Casio, and Timex manufacture some watches that sell for around $10. In *Double Prices,* trips are most akin to a luxury good, with high-end destinations and high-end hotels. For trips, customers almost always chose the higher priced option. They likely assume quality and prestige and, with that, a higher price. With the right anchoring, consumers can often be convinced that the good or service is deserving of a higher price.

FLIP FLOP

Flip Flop would appear to be more difficult than *Coming or Going, Double Prices,* or *Side by Side* because of the choices to flip, flop, or flip and flop the numbers to guess the price of the prize. However, the win rate of 60% (31 out of 52) in *Flip Flop* was almost as high as the 63% average win rate across the other three games.

Given three pricing possibilities, one might expect a sub-50% win rate. But the right answer was never to flip and flop, and contestants seemed to catch on, choosing to flip and flop only four times. Moreover, the two choices for the first two numbers (as shown, or "flipped") were typically $2,000 apart, creating a clear flip or non-flip (hence, flop) choice for the contestant. Flipping was the correct action in twenty-three games and flopping in the other twenty-nine games. The contestant flopped twenty-eight times, flipped twenty times, and flipped and flopped four times. Excluding the four instances of flipping and flopping, the win rate would have been 65% (31 out of 48).

In *Flip Flop*, the contestant went with the higher first two numbers half the time, and the higher choice was the correct pick in thirty-two out of fifty-two games. Contestants seemed to avoid the overestimation bias seen in *Coming or Going, Double Prices*, and *Side by Side*. I attribute this to the contestant not being able to visualize the price choices simultaneously and being slightly anchored by the first two digits displayed. The displayed number had the lower first two digits in exactly half the iterations of the game.

Key Tip

Never flip and flop. If you believe the first two numbers are correct, flop.

SQUEEZE PLAY

The sound effects of the numbers squeezing in *Squeeze Play* makes for an enjoyable acoustic experience. *Squeeze Play* is analogous to *Flip Flop* in that there are three possible price choices for a four-digit prize, but the right pick was never to squeeze out the fourth number, reducing the game to two choices. *Squeeze Play* was played fifty-seven times in seasons 47 and 48, the most of any pricing game, and won 60% of the time, with a $8,300 median prize value and $4,550 median earnings. Sometimes *Squeeze Play* was played for a five-digit prize, leaving the contestant with four possible numbers to remove from a six-digit string. Contestants won thirty of forty-six games (65% of the time) when playing for the typical four-digit prize and were four for eleven when playing for a five-digit prize, which had a higher degree of difficulty.

Not once was the right move in *Squeeze Play* to remove the next-to-last number in the string. Contestants caught on to this, removing that number only four times, all in four-digit prizes, which explains the high win rate versus random chance (i.e., 1 in 3). I believe that the showrunners avoid making the correct "squeeze play" the next-to-last number in the string so as to improve the game's odds, just as they avoid flip and flop as a correct choice in *Flip Flop* to improve that game's odds.

For four-digit prizes, the choice comes down to whether the second number or third number in the string is the second digit in the prize. Four-digit *Squeeze Play* therefore should be a bit more difficult than *Flip Flop*, in which the contestant is essentially guessing the first digit in the prize. The correct pick was to squeeze out the second digit twenty-two times and the third digit twenty-four times, whereas contestants squeezed out the second digit twenty-four times and the third digit eighteen times, along with four fourth-digit miscues.

For five-digit prizes, the 36% win rate (4 out of 11) was close to random chance if one assumes that removing the second, third, and fourth number are equally likely outcomes, and removing the fifth number is never the correct answer. Interestingly, in seven of these eleven games, the correct answer was to remove the second number. The contestant chose to squeeze out the second number only twice, winning each time. So with a five-digit prize, the contestant is well-served to first decide whether he thinks the second number is correct. If so, remove the third or fourth number. If not, remove the second number.

Key Tip

Never squeeze out the fourth number in a five-digit string or the fifth number in a six-digit string.

BALANCE GAME

Balance Game entertains viewers with the dramatic effect of the money bags balancing—or not balancing—with the prize. The contestant has three possible choice configurations from among the three bags: the two bags with the middle and the highest values, which sum to the highest total value; the two bags with the highest and the lowest values, which sum to the middle value; and the two bags with the middle and the lowest values, which sum to the lowest total value. The 42% win rate (15 out of 36) is evidence that contestants often had difficulty eliminating even one of the three options, although the median prize of $10,300 compensated for the degree of difficulty.

In thirty-six iterations of the game, the correct choice was the middle value fifteen times, the lowest value ten times, and the highest value eleven times. Contestants gravitated toward the middle value, choosing it nineteen out of thirty-six times, while choosing the lowest value ten times and the highest value seven times. Win rates were roughly 40% regardless of whether the contestant chose the lowest value, the middle value, or the highest value, suggesting *no* tendency to err more in one direction. Only three times did the contestant choose the lowest or highest value when the correct price was on the opposite end of the spectrum.

Errors were typical judgment errors. For example, on October 25, 2018, a player guessed $10,833 for a six-night trip for two to British Columbia from among $2,000, $3,000, and $8,000 bags, with an $833 residual. The retail price was $5,833, reflecting the short flight from Los Angeles.

The win rate was similar for prizes priced more than $10,000 (7 out of 18) and for prizes priced less than $10,000 (8 out of 18), as positive chance offset a more difficult setup. For more expensive prizes, price choices were much closer together relative to the prize value. The difference between the lowest choice and the highest choice was 58% of the actual retail price ($4,667) for prizes less than $10,000 and 34% ($4,278) for prizes in excess of $10,000. *Balance Game* was played twice for a car and won once when the price choices were farther apart.

Key Tip

The low- and high-priced bags (i.e., the middle option) are more likely to be the right choice. Favor the middle option, but pivot to the cheaper set of bags or the more expensive bags as circumstances suggest.

FREEZE FRAME: THE TROJAN HORSE

Freeze Frame looks harder than it is, given eight possible combinations for the price of the prize. The 40% win rate (14 out of 35 iterations) reflected the game's difficulty, and the $7,450 median prize value hardly compensated, leading to a sixty-ninth median earnings ranking. In *Freeze Frame*, there were always four pairs of two digits that were fifty or higher, for a $5,000-plus prize value (Figure 14.2). The minimum prize value across all pricing games is $5,000. Three times the player chose a sub-$5,000 price and lost.

Performance in *Freeze Frame* suffered due to underestimation. This is surprising given that the first two digits in the price were the most expensive option only five times, versus fourteen times for the second-most-expensive option, eleven times for the third-most-expensive option, and five times for the fourth-most-expensive option. The contestant guessed too low in seventeen of twenty-one losses, including three sub-$5,000 guesses, and too high in only four of twenty-one losses, as seen in Figure 14.3.

How do we make sense of the underestimation in *Freeze Frame*? As previously noted, players seem to underestimate prices in *TPIR* unless they are shown realistic pricing options. In *Freeze Frame*, four out of the eight options were always less than $5,000. These options seemed to orient the contestant lower, even if he chose from only the $5,000-plus options. The lower priced options functioned as a Trojan horse, distorting how players judged the price of the prize.

A successful heuristic I noticed in *Freeze Frame* is as follows: choose the two-digit number no less than 50 whose closest neighbor is further away, excluding neighbors less than 50. For the highest option, consider the difference versus the second-highest option, and for the second-, third-, and fourth-highest options, consider the minimum difference with the two adjoining numbers.

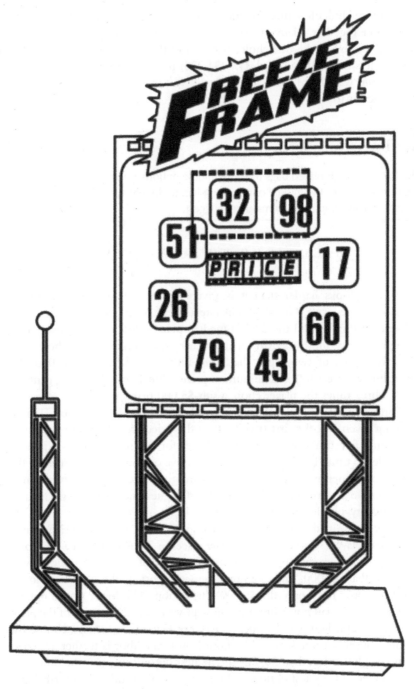

Figure 14.2 *Freeze Frame. Illustration courtesy of Omar Faruk*

Price Rank - Actual Price

		1st	2nd	3rd	4th	Total
	1st	1	1	1	0	3
Price Rank -	2nd	1	6	1	0	8
Player Choice	3rd	2	6	4	1	13
	4th	0	1	4	3	8
	Sub - $5k	1	1	1		3
	Total	5	14	11	5	35

Too Low - 17	Right Pick - 14	Too High - 4

Figure 14.3 *Freeze Frame* Performance by Price Rank

If the highest and second-highest options are tied under this algorithm, go with the second-highest option.

For a six-night trip to Quebec offered on June 28, 2019, and illustrated in Figure 14.2, 79 was the two-digit number whose closest neighbor, 98, was farthest apart. Although 98 is also nineteen units from 79, the tie goes to 79 because it is also nineteen units from the third-highest choice. Such a strategy would have won 71% of the time (25 out of 35) in seasons 47 and 48. *Freeze Frame* is plenty hard. One way for *TPIR* to make it easier is to *not* make the contestant choose the first two digits from numbers close together.

Key Tip

Focus on the $5,000-plus option whose first two digits have their closer neighbor farther apart. Pick that option unless confident otherwise. Never pick a sub-$5,000 option.

Trojan Horse

The term Trojan horse is generally associated with military strategies and computer viruses, in which tricking the adversary is common. In business, private equity firms are known to underinvest in their portfolio companies to inflate margins ahead of a sale. They hope to convince a buyer that the business can be run at higher profit margins and thus pay an inflated price for the business. The high and growing margins are a Trojan horse. Not only are the margins unsustainably high, but the buyer often has to invest more to make up for prior underinvestment. Low capital expenditure and research and development spending ahead of the sale process are often telltale signs.

PICK A NUMBER

Pick a Number had a 56% win rate (23 out of 41) and a median prize value of $8,200, leading to median earnings of $4,600. Nine times *Pick a Number* was played for five-digit prizes, including three cars, resulting in a higher average prize value of $9,450. All but two of the four-digit prizes involved picking the first digit in the prize, and all five-digit prizes involved picking the second digit. Contestants were five for nine when picking the second digit in a five-digit prize and seventeen for thirty when picking the first digit in a four-digit prize, for a similar 55%-plus win rate.

In seventeen out of thirty instances when picking the first number, the price was the middle choice, and in thirteen instances it was the high choice. *The low choice was never the price, even the two times the low digit was 6 and not 4 or 5.* Contestants seemed to grasp this, picking the middle number twenty times, the high number nine times, and the low number only once. It should therefore come as no surprise that the win rate when picking the first number was north of 50%.

The lowest number was also never the correct choice for the second digit in five-digit prizes. Here the contestant avoided the lowest option all nine times. The car setups did prove difficult, however. Car choices were $1,000 to $2,000 apart, and in the context of prizes that were two to three times more expensive than in a typical *Pick a Number* game. On the Christmas special in season 47, the contestant had to choose between $21,194, $23,194, and $25,194 for a Toyota Prius. He chose $25,194, and the price was $23,194. Contestants were much more successful when the pricing options were farther apart, as occurred on June 12, 2018, when the contestant played for a six-night trip to Buenos Aires. The choices were $10,600, $13,600, and $16,600. The contestant picked $13,600 and won.

Key Tip

Lean toward the middle option but do not hesitate to choose the high option if the prize seems on the expensive side. When picking the first number, never pick the low option.

PUSH OVER

Push Over functions like an easier version of *Freeze Frame*. There are fewer realistic choices, although the blocks are irrecoverably pushed over, preventing the contestant from altering his initial selection. *Push Over* was won 73% of the time (36 out of 49 games), with a $7,650 median prize value and $5,850 median earnings, ranking it twenty-third among *TPIR* pricing games and fifth among price guess games.

Why is *Push Over* won with such high frequency? Because most of the options can be eliminated. There are only five total choices, numbers 1–4, 2–5, 3–6, 4–7, and 5–8, as one block must be pushed over (Figure 14.4). Nor was the right answer ever pushing over all five blocks and leaving the leftmost numbers (i.e., numbers 1–4). That limits the choices to four, or three if playing for a five-digit prize.

For four-digit prizes, eliminating choices of less than $5,000 would have reduced the game to three choices in nine contests, to two choices in twenty-four

Figure 14.4 *Push Over. Illustration courtesy of Omar Faruk*

contests, and to one choice in seven contests. So it should come as no surprise that players won 65% of the time (27 out of 40) in four-digit *Push Over*. Moreover, that win rate would have measured 82% (27 out of 33) without five sub-$5,000 guesses and two instances of pushing over all five blocks.

It is generally not too difficult to choose between two choices of more than $5,000, because the choices are usually a few thousand apart. On June 10, 2019, there were two $5,000-plus choices for two ATVs: $9,543 and $5,438. Guessing the correct $5,438 price for the ATVs, as the contestant did, was straightforward, as ATVs do not cost $5,000 apiece.

Even with three feasible choices, as occurred in nine contests, one of the choices could usually be eliminated with some thought. Consider a six-night trip to Zion National Park in Utah offered on November 8, 2018. The three feasible choices were $5,768, $7,684, and $6,843. Utah is less than a two-hour flight from Los Angeles, and national parks often have less expensive lodging. Thus $7,684 was very unlikely. The contestant chose $5,768 and won.

Similar to *Freeze Frame*, the nonviable lower priced options were a Trojan horse, with ten out of thirteen losses involving the player choosing a price too low, including the five picks less than $5,000.

There were even fewer realistic choices in five-digit *Push Over*, which enabled a perfect nine wins in nine tries. In five games there were two realistic options, and in four games there was only one realistic option.

In summary, not only was the 73% win rate in *Push Over* deserved, but it easily could have been higher.

Key Tip

Never pick a sub-$5,000 option, and never pick the four or five leftmost numbers. Think before you push over. Often there is only one feasible choice.

CLOCK GAME

Clock Game is a game of quick thinking and dexterity with numbers, rather than a game of strategy or heuristics. The contestant should move higher or lower in $50 or $100 increments from his initial guess, and then move in smaller increments to home in on the price of the prize. For example, if the prize costs $743, the contestant might start at $900, go to $800 if told "lower," then $700 if told "lower" again. When told "higher," he might go to $750, and then $725 when told "lower." He could then try $730, $735, and $740, each time being told "higher," before being told "lower" at $745. He would then go down to $744 and $743, the correct price.

Contestants won eight out of thirteen times (62%), played for a median prize value of $10,400, and had median earnings of $6,650. Approaching *Clock*

Game as described here should allow the contestant to win with a few seconds to spare. In contrast, contestants moving in odd increments or forgetting the high/low pricing history can lose valuable time and come up short. Only once did the contestant not succeed in pricing one of the smaller prizes in the allotted time.

Contestants winning did so in an average of 23.25 seconds, spending 14.5 seconds on the first prize and 8.75 seconds on the second prize after becoming more adept at the game. The fastest two wins were in 17 seconds. Never were the prizes priced more than $1,000 or less than $500.

Key Tip

Start with sizable incremental moves and home in on the price using smaller incremental moves. Do not move so fast as to throw off your concentration or memory.

SAFE CRACKERS

In *Safe Crackers*, the contestant must guess the price of a three-digit prize in order to win a larger prize. Contestants won *Safe Crackers* in twelve out of eighteen games with a median prize value of $9,500. Every episode of the game had a zero in the number, and zero was the last digit in all eighteen iterations of the game. Contestants, to their benefit, always put zero last. With zero as a known third digit, the game basically comes down to the contestant choosing which of two non-zero numbers is the first digit in a three-digit prize and which of two non-zero numbers is the second digit.

Both non-zero number options in *Safe Crackers* were 5 or higher, except for three games in which one of the two non-zero numbers was a 4. In each of those three cases the three-digit prize started with the higher number. The three-digit prize was the higher priced option in twelve of eighteen games, although that included three games with four as an option. Contestants guessed the higher priced option ten times.

In all but four iterations of the game the two non-zero numbers were only two units apart (~$200), creating a difficult setup and limiting wins to eight out of fourteen tries. On September 25, 2018, the contestant lost when guessing $970 rather than $790 for a pair of sunglasses. The contestant won all four times when the two non-zero digits were three or four units apart, equating to approximately a $300 or $400 price difference.

Key Tip

Always choose zero for the last number. If the digit 4 is an option for the first number, do not choose it. Lean toward the higher priced option but be flexible.

---

2 FOR THE PRICE OF 1

2 for the Price of 1 is similar to *Safe Crackers*, except that in *2 for the Price of 1* the contestant has different number choices for each digit and is given one free digit. Zero was an option for the third digit in virtually every game and was always the third digit when presented as an option. The one time that zero was not an option was when a contestant played for a Volkswagen Jetta on May 13, 2019, which resulted in a loss. Excluding this contest, contestants won six out of fifteen times and played for an $8,700 median prize. Notably, six of the nine losses involved choosing a non-zero number as the third digit in the three-digit prize.

It does not require a math degree to realize that contestants should ask for a free second digit, choose zero for the third digit, and use their judgment to pick the first digit. Contestants were six for nine when they followed this strategy. On June 24, 2019, the contestant played for a six-night Australian vacation by guessing the price of a surfboard whose first digit was 3 or 7, whose second digit was 1 or 5, and whose third digit was 0 or 6. The contestant asked for the second digit, picked 0 for the third digit, and smartly went with 7 as the first digit, winning the surfboard and the trip, worth a combined $10,540.

Fortunately, contestants were mostly presented with easy choices for the first digit. Never was the first digit of the three-digit prize below 5, even though number choices below 5 were presented in nine out of sixteen iterations of the game. Choosing the higher first digit would have been correct twelve out of sixteen times, although only three out of seven times when both first digit choices were 5 or more.

> ## Key Tip
>
> A strategic contestant should be victorious more than 75% of the time. Simply ask for the second digit, pick 0 for the third digit, and choose the higher first digit if presented with a choice below 5. If not presented with a first digit choice below 5, use your best judgment.

BONKERS

Bonkers is bonkers. The fast pace of the game often throws off contestants. The 53% win rate (9 out of 17) is just fair against a modest $7,600 median prize, while median earnings of $4,000 ranked fifty-eighth.

In *Bonkers* there are $2 \times 2 \times 2 \times 2 = 16$ choices, but only $2 \times 2 \times 2 = 8$ options conditional on knowing the first digit with directional certainty. And it is usually clear-cut as to whether the first displayed digit is too low or too high. In fifteen out of seventeen games the number displayed for the first digit was 5 or lower, and all fifteen times the correct choice was higher. This should not come as much of a surprise, as pricing game prize values were never less than $5,000. Still, three out of seven losses when the first displayed digit was 5 or lower involved the player trying at least one combination with a lower first digit, using up precious time. On an episode airing February 18, 2019, the contestant played for a sectional couch and a seventy-inch smart 4K TV with a displayed price of $5,673. He guessed a lower first digit on four of his six choices, causing him to run out of time and lose.

Realistically, the contestant can cycle through as many as six choices in thirty seconds. In the eight losses, the contestant cycled through three choices three times, four choices once, five choices twice, and seven choices twice. Therefore, if the contestant has certainty on the first digit, he should have at least a 3/4 chance of winning just by trying six of the eight possibilities for digits 2–4 before the thirty seconds expire. This did not prevent a frustrating loss on November 1, 2019, when the contestant played for a seven-night Aruba vacation. The displayed number was 5,347, and the contestant tried six different combinations with the first digit greater than 5 (and a seventh try that repeated a guess), but not the winning combination of $9,754.

Of course, the contestant has little visibility on the second digit, let alone the third and fourth digit. Still, one can surmise that for displayed numbers closer to nine, the correct number in the price of the prize is likely lower, and for displayed numbers closer to one, the correct number is likely higher. Why? Because if the number is three, there are six number choices above but only three below. A randomly distributed number would thus be two-thirds likely to be above three and one-third likely to be below three.

Let us formalize this approach. In the first choice and all successive choices, choose a higher first number if the displayed number is 1–6 and a lower number if the displayed number is 7–9.* Choose the second, third, and fourth digits to maximize the number of digits above or below the number shown. If incorrect, flip the selections for displayed numbers 4 and 5, trying one digit at a time. Note the digit 5 should lead to a lower initial choice as there are five lower digits, 0–4, and four higher digits, 6–9.

My approach would have resulted in wins 100% of the time during seasons 47 and 48 in one or two tries, and all but once on the first try. Consider the 4,736 string on January 28, 2019, for a six-night trip to the Dominican Republic. My strategy suggests placing the disks higher for the first number,

---

*The first number in the number string was never 6 during seasons 47 and 48. The ten times it was 6 during seasons 45 and 46, the right choice was always higher.

lower for the second, higher for the third, and lower for the fourth. That was indeed the correct set of choices for the $6,164 trip. Only on April 8, 2019, would the contestant not have won on his first attempt. Here a six-night trip to Boston priced at $7,970 versus the 4,357 string. The third digit, 7, was above the number 5 in the string, whereas my approach suggests initially guessing lower for the third digit. But the contestant still would have won on his second guess with ample time left after changing his choice for the third digit to higher.

I have developed a foolproof backward-looking strategy to win at *Bonkers*. Now that's bonkers.

Key Tip

For digits 2 through 4, the first guess should be higher or lower based on whether there are more numbers higher or lower than the displayed number. For the first digit, go higher against 1–6 and lower against 7–9. If unsuccessful, alter one's guess for the digit(s) whose displayed number is 4 or 5. Continue altering one's guess for digits closest to 4 or 5, one at a time, until victorious.

CHECK GAME

Check Game is very winnable with a simple heuristic. Yet contestants won only six out of sixteen contests, with a median prize value of $8,050 and median earnings of $3,000, ranking sixty-sixth. The target range for the value of the check and the prize increased from $7,000 to $8,000 in season 47 to $8,000 to $9,000 in season 48.

In nine of the ten instances where the contestant lost, he wrote a check that put the total value (check plus prize) above target range. Even after increasing the range to $8,000 to $9,000, two of the three losses involved the contestant writing a check too large. With the *Check Game* prize always falling between $5,388 and $6,543, the player would have won ten out of sixteen times by always writing a $1,500 check during season 47 or a $2,500 check during season 48.

During season 47, the check-plus-prize value averaged nearly $8,500, $1,000 more than the midpoint of the $7,000 to $8,000 range. During season 48, the average check-plus-prize value totaled $8,700, still skewing modestly high. Across both seasons, the bias was most noticeable for non-trip prizes, which contestants won only once in seven tries, versus trip prizes, in which they were six for ten. In *Check Game*, underbidding took the form of underestimating the prize value and compensating by writing a check too large.

> ## Key Tip
>
> A check of $2,500 should position the contestant to win about two-thirds of the time. Start at $2,500 and adjust upward or downward if the prize appears materially above or below $6,000.

RANGE GAME: STRATEGIC PATIENCE

Most fans love *Range Game*. There's the noise of the moving range. The contestant frets about stopping the range finder too soon or not soon enough, all the while the audience yells out what to do. *Range Game* is one of the more psychologically trying games on *TPIR*, yet it is easier than it appears, and nobody should be surprised at the 72% win rate (26 out of 36). Couple this with a $10,200 median prize value, and the median earnings of $7,050 placed it a solid twenty-first among pricing games.

Range Game seems difficult because a $600 range for an item typically priced between $7,000 and $15,000 is not a wide range. It is 4 to 9% of the prize value, and closer to 4% for more expensive prizes. But *Range Game* was relatively easy in that the price never once fell in the bottom $200 or top $200 of the $600 range. If the contestant had followed the strategy of never stopping the range finder to cover prize values in the lower $200 or upper $200 of the range, he would have had a 75% chance of winning, as the $150 range finder would fall somewhere in the middle $200 band.

If the contestant had stopped the range finder exactly in the $225–$375 band relative to the $600 range, he would have won all but three iterations of the game. The only instance in which the prize was priced at exactly $400 in the range was during Big Money Week on October 10, 2018, when a Chevy Cruz sedan was being awarded. The car priced at $20,400 relative to the $20,000–$20,600 range, a more difficult setup than typically seen. The contestant stopped the range at $20,240–$20,390 and just missed a win. On average, the prize value fell $6 above the midpoint of the $600 range. In fifteen of the thirty-six iterations of the game, it fell between the $280–$320 mark of the $600 range.

Contestants were inclined to stop the range finder too early. The average range, relative to the $600 interval, was $196 to $346. I attribute this to the psychological challenge of letting the slow-moving range finder move the necessary distance rather than to underestimation, especially since the $600 range is small relative to the prize value. Nine of the ten losses involved stopping the range prematurely, and only one loss involved stopping the range too late. Moreover, two of the premature stops were very premature. On June 12, 2019, when playing for a South African safari, the contestant stopped the range only $60 from the bottom, as illustrated in Figure 14.5. And on April 10, 2020,

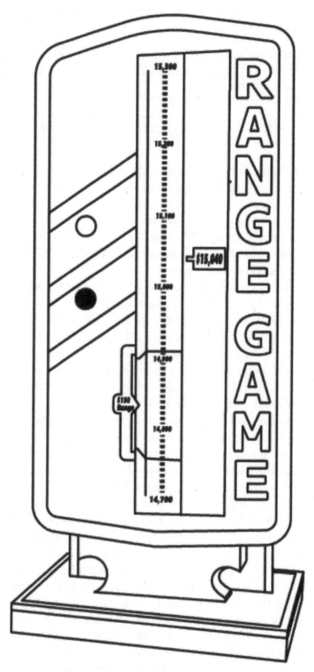

Figure 14.5 *Range Game. Illustration courtesy of Omar Faruk*

when playing for a luxury spa and hammock, the contestant stopped the range only $120 from the bottom.

So hope to play *Range Game* if you get called out of Contestant's Row, and play it with patience. You have an excellent chance of winning a nice prize, often in excess of $10,000!

Key Tip

Never let part of the range finder fall in the lower $200 or upper $200 of the range. Stop exactly in the middle ($225–$375) unless you have a strong inclination to go slightly higher or lower.

Strategic Patience

Patience is a virtue, particularly in negotiations. Consider buying a house. A great way to get a good deal on a house—if not amid a tight market and if not in a rush to move—is to wait for the list price to be lowered. If worried about the house being sold to another buyer, just let the selling realtor know of your interest, and he will reach out if there is buyer interest. But waiting causes anxiety, which often causes buyers to make an offer too early. Typically, the seller is going to show less price flexibility during the first couple of weeks their homes is on the market. They think they've got something special. More often than not, though, the property is priced too high. After a while the seller will explicitly lower the list price or signal a willingness to entertain lower offers.

15

Cash Games

Cash games are contestant favorites on *The Price Is Right* (*TPIR*). Not much is desired more by contestants than cold, hard cash, as physical prizes might not be to their liking. Cash games were played only 243 out of 356 episodes, or slightly more than two times in every three episodes, and only twice on a single episode. *Plinko* is the most popular and most played cash game, and *Grand Game* was also played quite frequently.

Most cash games have large potential cash prizes but uninspiring median win values, as many contestants do not win the full cash prize. Median earnings were $4,850, but a more representative $4,400 when excluding *Pay the Rent*, as seen in Figure 15.1. In *Pay the Rent*, the results are skewed by a $100,000 win on the season 47 opener, when the game had an unusually easy setup to facilitate a big win. Without this "facilitated" win, contestants would have averaged $5,300 in *Pay the Rent* across fifteen contests rather than $11,300 across sixteen contests.

Game	Played	Wins	Losses	Win Rate	Average Prize	Median Prize	Average Win	Median Win	Min Win	Max Win	Median Earnings	Earnings Rank
Grand Game	40	12	28	30%	$ 13,000	$ 10,000	$ 10,000	$ 10,000	$ 10,000	$ 10,000	$ 3,125	65
Half Off	34	9	25	26%	$ 12,599	$ 11,397	$ 15,464	$ 11,286	$ 10,308	$ 51,594	$ 3,243	64
Hot Seat	24	11	13	46%	$ 23,652	$ 20,314	$ 8,863	$ 10,173	$ 5,140	$ 20,210	$ 4,963	42
It's in The Bag	25	6	19	24%	$ 18,560	$ 16,000	$ 12,000	$ 12,000	$ 8,000	$ 16,000	$ 3,600	60
Pay the Rent	16	10	6	63%	$100,000	$100,000	$ 18,000	$ 10,000	$ 5,000	$100,000	$11,250	2
Plinko	47	24	23	51%	$ 59,411	$ 50,194	$ 20,603	$ 12,190	$ 10,060	$202,165	$ 7,054	20
Punch-A-Bunch	29	7	22	24%	$ 25,810	$ 25,288	$ 5,944	$ 5,273	$ 5,142	$ 10,131	$ 2,378	73
Time is Money	28	9	19	32%	$ 20,000	$ 20,000	$16,958	$18,914	$12,383	$ 20,000	$ 5,638	34
Total/Wtd Avg	243	88	155	36%	$ 31,608	$ 28,508	$14,743	$11,422			$ 4,867	
Ex-Pay The Rent	227	78	149	34%	$ 26,788	$ 23,469	$14,325	$11,605			$ 4,417	

Figure 15.1 Statistical Summary—Cash Games

- *Grand Game*: The contestant is shown six grocery items and a limit price ranging from $3.50 to $8.50. He selects the four grocery items that are less than the limit price to win $10,000. The contestant begins with $1. The first correct pick gets the contestant from $1 to $10, the second pick to $100, the third pick to $1,000, and the fourth and final pick to $10,000. The contestant can stop at $1,000, rather than playing for $10,000 and potentially losing everything.

- *Half Off*: The contestant plays for $10,000 in cash. There are sixteen boxes numbered one through sixteen on a tiered table. One has $10,000 cash in it. The contestant can eliminate half the boxes at a time by correctly guessing which of two small items (usually priced between $20 to $200) is half off. He can do this three times with three sets of items. If he gets all three small item guesses correct, he wins a $1,000 bonus. He then chooses which of the remaining boxes—from the sixteen, eight, four, or two choices—has the $10,000 cash inside. He wins $10,000 if correct.

- *Hot Seat*: The contestant plays for up to $20,000 in cash. He is shown five small items (ranging from $10 to $200) and five displayed prices. He then has thirty-five seconds in the "hot seat" to choose whether the actual price for each of the five small items is higher or lower than the displayed price. Next, he is shown small items for which he has made the correct call (higher or lower). The first correct pick is worth $500, the second $2,500, the third $5,000, the fourth $10,000, and the fifth $20,000. The contestant loses everything with an incorrect pick, but he can stop at any point. He gets to see which item is to be revealed next before deciding whether to stop and bank the cash or continue play.

- *It's in the Bag*: There are five grocery bags, each with a price that corresponds to one of six grocery items. The contestant selects which item goes with each grocery bag/price, with one item remaining unused. The first correct pick wins $1,000, the second $2,000, the third $4,000, the fourth $8,000, and the fifth $16,000. The contestant can stop at any point and loses everything with an incorrect pick. The bags/prices are usually sequenced so that the first is the lowest, the second the highest, the third the second lowest, the fourth the second highest, and the fifth is the midpriced item.

- *Pay the Rent*: There is a house with four levels. The contestant places six grocery items in the four levels of the house: one in the mailbox (lowest level), two on the first floor (second level), two on the second floor (third level), and one in the attic (fourth level), such that the price of the item(s) on each floor is successively higher. The mailbox is an automatic $1,000, the first floor $5,000, the second floor $10,000, and the attic $100,000. The player can stop at any point but loses all if the next floor up is less expensive than the one below.

- *Plinko*: The contestant is given one *Plinko* chip to start and can win up to four additional *Plinko* chips by guessing whether an incorrect displayed price for each of four small two-digit prizes ($10–$99 in value) has the correct first number or the correct second number. Each chip is placed at the top of the *Plinko* board, which has nine slots. The $10,000 slot is in the center, with a $0 slot on each side of the $10,000 slot, and working outward, two $1,000 slots, two $500 slots, and two $100 slots.

- *Punch-a-Bunch*: There is a punchboard that conceals a paper slip with a dollar value in each of its fifty paper-covered holes. The cash slips typically range from $100 to $10,000. The contestant can win up to four punches by deciding if the price of each of four small items is higher or lower than the price shown. He then punches one, two, three, or four paper boxes on the board. Drew picks one box, removes the cash slip, and reveals the cash prize. The contestant can quit with the cash prize or turn in the cash slip and play for a better prize with his remaining punches, stopping at any time. If he plays through the last punch, that cash slip becomes his cash prize value.

- *Time Is Money*: The contestant has ten seconds to place each of five grocery items on a $0–$2.99 table, a $3.00–$5.99 table, or a $6.00 and more table. If correct, he wins $20,000. If incorrect, he has thirty seconds to rearrange the items one or more times while the $20,000 prize meter reduces to zero ($666 per second).

GRAND GAME

Grand Game is the second-most-played cash game. It has mediocre results, with a 30% win rate (12 out of 40) and $3,150 median earnings, ranking sixty-fifth among price games.[*] There are no strategies or heuristics for better play in *Grand Game*, but one can probabilistically analyze what results to expect. With four of the items less than the limit price, if the contestant is guessing at random, he would have a 4/6 chance of guessing a first grocery item below the limit price. Conditional upon continued success, he would have a 3/5 chance of guessing the second item correctly, a 2/4 chance on the third item, and a 1/3 chance on the fourth item. Thus, a random set of guesses would win $10,000 about 7% of the time. $\left(\frac{4}{6} \times \frac{3}{5} \times \frac{2}{4} \times \frac{1}{3} \right) = \frac{1}{15}$ or 6.7%.

Given that some grocery items are clearly priced below the limit price, I estimate that the probabilities run more like 100% for the first pick, 90% for the second pick, 80% for the third pick, and 50% for the fourth pick. These

---

[*]Results are inclusive of a $5,000 partial win when the game was played for $50,000 during Big Money Week and the contestant stopped short of playing for the grand prize.

probabilities suggest a 36% win rate: $\left(1 \times \frac{9}{10} \times \frac{8}{10} \times \frac{1}{2}\right) = 36\%$. This is slightly better than the 30% observed, likely a function of chance. During seasons 47 and 48, one contestant got eliminated on his first pick, three on their second picks, four on their third picks, and nineteen on their fourth picks, with one player stopping after his third pick at $5,000 during Big Money Week when playing for $50,000. No other contestant stopped at $1,000 after a third successful pick and rightly so. One's chances on the fourth number are no worse than a one in three, whereas the cash prize increases by a factor of ten.

Consider the varying difficulty of the different grocery items. On June 20, 2019, the limit price was $3.50; the $0.55 cat food was obvious, as were the string beans priced at $1.39. The energy drink was the contestant's third pick. Although its price was higher than I expected at $2.79, it still fell below $3.50 target price. The final pick was among crackers, Aquaphor lip treatment, and hash browns, but really between crackers and hash browns, since Aquaphor costs well more than $3.50. The contestant went with crackers, priced at $4.19, instead of the hash browns, priced at $2.99, and lost. This game came down to a fourth pick between two items, as is the case for most iterations of *Grand Game*.

In *Grand Game*, the least expensive prize and the second least expensive prize always fell well below the limit price. The third least expensive prize averaged about $1.50 less than the limit price. The fourth least expensive prize often fell less than $1.00 below the limit price and $1.50 to $2.00 below the second most expensive prize. It is here where the contestant usually faltered in his quest for $10,000. In fact, of the thirty-one fourth and final picks, twenty-three of them (in hindsight) involved the contestant trying to pick the fourth least expensive prize. Contestants were right in only eight of these twenty-three tries, suggesting a counterintuitive setup for the fourth least expensive prize. This notion is supported by the fact that contestants did not win more when the price difference between the fourth least expensive prize and the second most expensive prize was more than $2 versus less than $2, as seen below.

Result	Win	Loss	Stop Early
< $2 delta, fourth least expensive item and second most expensive item	6	14	1
> = $2 delta, fourth least expensive item and second most expensive item	5	13	0

HALF OFF

Half Off is one of my least favorite games, because the outcome depends almost entirely on chance. The game's low win rate, nine out of thirty-four (26%), reflects the rarity of getting all three small items correct and being able to select from two boxes. Detailed performance in *Half Off* was as follows.

- Nine times (26%) the contestant reduced the boxes down to two choices and won six out of nine times.
- Fifteen times (44%) the contestant reduced the boxes down to four choices and won three out of fifteen times.
- Eight times (24%) the contestant reduced the boxes down to eight choices and failed to win.
- Two times (6%) the contestant chose from all sixteen boxes and failed to win.

Chance would have suggested a similar 9.4 wins based on the distribution of eliminated boxes as shown in the preceding list: $\left(9 \times \frac{1}{2}\right) + \left(15 \times \frac{1}{4}\right) + \left(8 \times \frac{1}{8}\right) + \left(2 \times \frac{1}{16}\right) = 9.4$.

Contestants correctly guessed the half-off item 65 times out of 102 attempts in 34 games, averaging 1.91 qut of 3 tries (64%). Per the binominal formula, 64% success in guessing the half-off item would lead to the following expected outcomes.

- Two boxes left 26% of the time: $0.64 \times 0.64 \times 0.64 = 0.26$
- Four boxes left 44% of the time: $0.64 \times 0.64 \times 0.36 \times 3 = 0.44$
- Eight boxes left 25% of the time: $0.64 \times 0.36 \times 0.36 \times 3 = 0.25$
- All sixteen boxes left 5% of the time: $0.36 \times 0.36 \times 0.36 = 0.05$

This is also very similar to what was observed during seasons 47 and 48.

Interestingly, contestants would have been successful in 70 out of 102 tries (69%) had they always chosen the item with the lower displayed price. This result is unlikely due to random chance. There is only a one in two thousand random chance of the lower priced item being half off in 70 of 102 trials, as calculated per the binomial distribution. Players chose the lower priced item 73 out of 102 tries (72%), so they partially caught on to this pattern.

> ## Key Tip
>
> Lean toward the less expensive item when choosing which item is half off for the purpose of eliminating boxes unless reasonably certain the more expensive item is half off.

HOT SEAT

Hot Seat is my personal favorite pricing game. It combines strategy, pricing knowledge, the pressure of the clock, and key decisions of whether to quit at $5,000 or $10,000 or play for the full $20,000. After the contestant spends thirty-five seconds in the hot seat, deciding whether displayed prices for small items are too high or too low, he is psychologically put in the hot seat, deciding when to bank his money and when to play for more. Contestants "won" *Hot Seat* 46% of the time (11 out of 24), winning $5,000, $10,000, or $20,000. The median win value was $10,200, given a $10,000 median cash win and the approximate $200 value of the smaller items. The average win value was a lower $8,900.

What makes *Hot Seat* tricky is that a contestant's correct picks are revealed before he is shown any of his incorrect choices. Supposedly correct choices are not shown in any particular order, although I wonder if *TPIR* reveals the "easier" picks first. Unless the contestant is certain that he has made more than two mistakes, he has no reason to stop before he has gotten three prizes correct and reached $5,000. Even a random set of guesses would result in three or more correct choices half the time. And the contestant should have better than 50/50 odds, particularly on easier items.

Consider the distribution of actual correct choices by contestants and when the contestant chose to stop. Figure 15.2 shows the number of correct choices in *Hot Seat*, regardless of when the contestant stopped. Twice contestants had a perfect showing, and five times they got only two choices correct, with the other twenty games resulting in three or four correct choices. Contestant

# of Correct Picks	0	1	Individual Trial Success Probability 66% 2	3	4	5
Actual Correct Picks, # of Games	0	0	5	9	8	2
% Correct, Share of Games	0.0%	0.0%	20.8%	37.5%	33.3%	8.3%
Binomial Probability	0.5%	4.4%	17.1%	33.2%	32.3%	12.5%
# of Times Stopped Correctly			2	3	4	1
# of Times Stopped One Too Early			0	0	2	1
# of Times Played on and Lost			3	6	2	

Figure 15.2 Successful Picks in *Hot Seat*

picks correctly identified whether the price of the prize was higher or lower than the displayed price 66% of the time ("The Success Rate").

The binomial distribution with a success rate of 66% is also shown in Figure 15.2, and the probabilities are not dissimilar from the results observed. Note that players had better success than the binomial distribution in getting two or three choices correct, similar success in getting four calls correct, and less success in getting five calls correct. Why? Because there are usually a couple of items in which the contestant is confident as to the price being higher or lower and a couple of items for which the player's choice is no better than a 50/50 guess.

Contestants generally made wise decisions about when to stop, also seen in Figure 15.2. Twice they stopped after two correct choices, and both times they had two choices correct. Only twice did they stop after three correct choices when they had four choices correct, and only once did they stop after four correct choices when they had all five correct. Three contestants lost on the third price revealed, and six lost on the fourth price revealed. Eight had four correct choices, with four correctly stopping at $10,000, two stopping at $5,000, and two playing for $20,000 and losing. Only two contestants had all five choices correct, with one winning $20,000 and one stopping at $10,000.

How far should contestants play in *Hot Seat*? They are usually better off playing for $10,000 versus stopping at $5,000 but better off stopping at $10,000 versus playing for $20,000. In ten of twenty-four iterations of the game (42%), the contestant had four or five correct choices, and in nine of twenty-four iterations of the game (38%), he had three correct choices. Conditional on the player getting three choices correct, he got four or five choices correct a touch more than 50% of the time. Had he always risked $5,000 to get $10,000, he would have won $10,000 10/19 of the time and ended up empty-handed 9/19 of the time, for an expected value of $5,250. Of course, some players will have a high degree of certainty as to whether the fourth item to be revealed has a higher or lower price than the displayed price. And some will not. Playing for $20,000, however, is not wise unless the contestant has near certainty regarding the fifth item. Upon reaching $10,000, he would have won $20,000 only two of eight times (25%).

The binomial distribution with a 66% success rate argues for a similar strategy, per Figure 15.3. There is a 57.4% chance of reaching $10,000 upon reaching $5,000 and a 27.9% chance of reaching $20,000 upon reaching

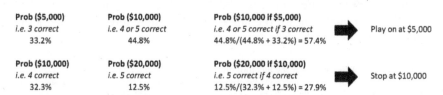

Figure 15.3 *Hot Seat* Play on Probabilities per Binomial Distribution

$10,000. Expected value analysis aside, a risk averse player may prefer $5,000 to a 57.4% chance of $10,000 and a 42.6% chance of zero.

How low would the success rate need to be for a contestant to stop at $10,000? And how high to justify playing for $20,000 upon reaching $10,000? A success rate of 60% or less argues for stopping at $5,000 after three correct calls, whereas a success rate of 84% or more argues for playing for $20,000 after four correct calls.[†]

In practice, the later prizes revealed are less likely to be called correctly (i.e., a success rate less than 66%) than the earlier prizes revealed (i.e., a success rate more than 66%), suggesting that playing for $10,000 is more of an expected value wash. Decisions should also be tied to the contestant's degree of confidence in calling the item in question. When contestants played for $10,000, they were correct eight out of fourteen times (57%), whereas three of five contestants who stopped at $5,000 would have lost all had they played for $10,000. Clearly a player's level of conviction regarding the item in question should affect his decision.

Key Tip

Always play for $5,000 unless you are confident about having misjudged the third item. Play for $10,000 if you are 60% confident in your choice for the fourth item. Do not play for $20,000 unless you are certain your choice for the fifth item (higher or lower) is correct.

IT'S IN THE BAG

It's in the Bag is a souped-up version of *Grand Game* and a bit harder to win, with a 24% win rate (6 out of 25). Wins tended to be larger, leading to slightly higher median earnings of $3,700 versus $3,100 for *Grand Game*. Players must place five grocery items in bags with specific prices rather than just calling out which items are less than a limit price.

Across twenty-five iterations of *It's in the Bag*, contestants stopped three times each at $2,000, $4,000, and $8,000 levels. They guessed incorrectly once on the $2,000 bag, three times on the $4,000 bag, and twice on the $8,000 bag. Ten times they went for the full $16,000, winning three times and losing seven times.

[†]With a 60% probability of success, the binomial distribution predicts three, four, and five correct choices with probabilities 34.6%, 25.9%, and 7.8%. Because 34.6% > (25.9% + 7.8%), the contestant should stop at $5,000. With an 84% probability of success, four and five correct choices occur with probabilities 39.8% and 41.8%. The contestant should thus play for $20,000.

Where did contestants falter? The third bag (i.e., $4,000) was called correctly nineteen times and incorrectly five times (including three games in which the contestant had already stopped at $2,000). In the nineteen games in which the contestant called the first three bags correctly, he called the fourth bag (i.e., $8,000) correctly thirteen times (including three games in which he had already stopped at $4,000). And the fifth bag (i.e., $16,000) was correctly chosen only four out of thirteen times after the first four bags were called correctly (including once in which the contestant had already stopped at $8,000). The odds for the fifth bag, assuming correct choices for bags 1–4, should be at least 50/50; in practice the fifth bag often featured a confusing grocery item or an item with an atypical price. In contrast, performance on the fourth bag was stronger than expected.

Atypical items for the fifth bag included a ten-ounce scented candle for a $4.99 bag on June 10, 2019, with the contestant instead guessing a salad kit. On May 21, 2019, the contestant picked granola for the fifth and final $4.19 bag when the correct pick was a sixteen-ounce container of horseradish sauce. And on January 29, 2019, the contestant chose shampoo for the final $4.99 bag when the right answer was crackers. In all three cases, the last bag was a toss-up and a difficult one at that.

The right strategy usually is to play for $8,000. Had every player stopped after the $8,000 bag or lost before, there would have been thirteen wins in twenty-five games and average earnings of $(13 \times \$8,000)/25 = \$4,160$, versus the $3,600 average observed. Stopping at $4,000 (or losing before) would have yielded lower average earnings of $(21 \times \$4,000)/25 = \$3,360$. And playing all the way to $16,000 would have led to four wins in twenty-five games, for average earnings of $(4 \times \$16,000)/25 = \$2,560$.

> ## Key Tip
>
> Contestants usually maximize expected earnings by playing until they win $8,000. Play until $8,000, unless lacking confidence in your third pick or being especially confident in your fifth pick.

PAY THE RENT

Pay the Rent is the game with the largest cash prize and for good reason. There were no $100,000 winners except for an unusually easy win on opening day season 47. Here the price of the most expensive item exceeded the combined value of any two of the remaining five items, vastly simplifying one's choices for the top two levels. Excluding that $100,000 win, total cash winnings of

$80,000 across fifteen contests was hardly notable. Seven contestants stopped at $10,000 and two stopped at $5,000, and none of those contestants would have won $100,000. Five contestants lost trying for $100,000, and one lost trying for $10,000.

How many possible combinations are there across the levels of the house, illustrated in Figure 15.4? Six choices for the first level, ten pairings for the second level (among the five remaining groceries), and three pairings for the third level (among the three remaining groceries), leaving one remaining item for the top level. In total, 180 combinations (6 × 10 × 3). The challenge in winning $100,000 is to price the top two levels correctly. Usual practice was choosing the second and third most expensive items for the third level and then failing on the top level if choosing to play on. It is easy, however, for the contestant to stop on the third level and win $10,000. Thirteen of fifteen contestants would have won $10,000 doing just that.

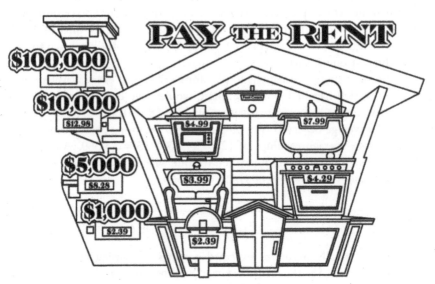

Figure 15.4 *Pay the Rent. Illustration courtesy of Omar Faruk*

The key to *Pay the Rent* is to choose items on the third level that are *not* the second and third most expensive groceries. Otherwise, the most expensive grocery item on the top level will almost always *not* be priced higher than the combined price of the grocery items on the third level. The results highlight this predicament. On October 5, 2018, the contestant chose inexpensive items for the second level, namely the $0.60 item and the $4.99 item. This left three items that could not be paired up to win $100,000 on the third and top levels, as they turned out to be priced at $2.29, $10.99, and $8.99. The contestant effectively needed to save the $0.60 item for the third level and pair it with the $8.99 item so that the sum total ($9.59) fell short of the $10.99 item on the top level. There was no other pairing of items on the top two levels that would have won $100,000, although the $2.29, $3.49, and $4.99 items could have been paired up any which way on the first two levels.

In nine of sixteen iterations of *Pay the Rent,* the player chose the most expensive item for the top levels and either the second and third, second and fourth, or third and fourth most expensive items for the second level. None were $100,000 winning combinations, except for the easy setup on opening day season 47. In the other seven games, he failed to choose the most expensive item for the top level, always a necessary condition for winning.

Choosing the least expensive and second most expensive items for the third level and the most expensive items for the top level was always a successful top two floor strategy to win $100,000 in *Pay the Rent.* With that said, four out of sixteen iterations of *Pay the Rent* had two ways to win, eleven out of sixteen iterations had three ways to win, and opening day season 47 had nine ways to win.

Let's explore this third level strategy further. On October 30, 2018, the contestant should have paired the $7.99 and $0.35 items on the third level, one floor below the $9.99 item, while pairing the $4.19 and $3.29 item on the second level and putting the $6.25 item on the bottom level. Here only one pairing worked on the bottom two levels. On June 3, 2019, the contestant should have paired the $2.39 item with the $7.99 item on the third level so that it totaled less than the $10.99 item on the top level. The $3.99, $4.29, and $4.99 items could have been paired all three possible ways on the bottom two levels while preserving the win. Notably, in all but two instances of the game, the bottom two levels would have been "solved" by putting the third most expensive item on the bottom level and the fourth and fifth most expensive items on the second level. This is a reasonable strategy for the bottom two levels.

Contestants could have done very well following my strategy. If they were able to identify the most expensive, the second most expensive, and least expensive prizes, they would have solved the top levels. In five games, this would have been enough to win $100,000, because any three combinations of the remaining items on the first two levels would have worked. Other setups were a bit more challenging. In six games only one of the three combinations would have worked on the first two levels, and in four games only two combinations.

If one can get the top two levels correct 50% of the time and the bottom two levels correct 60% of the time, then the odds of winning $100,000 with strategic play would measure 30% = 50% × 60%, for an expected win value of $30,000! What a fabulous setup! And a strategy not one contestant followed.

Unfortunately, thirteen out of sixteen contestants opted to put the least expensive item in the first level mailbox, not realizing its strategic significance on the third level. Given that the item placed in the first level mailbox only needs to cost less than the *two* items on the second level, it is an extreme over-reaction to the price ranking exercise to place the least expensive item on the first level.

I have focused on the winning solution(s) that involve pairing the second most expensive and least expensive items on the third level. But there are often other winning possibilities. In six games there was a winning solution that involved pairing the third and fourth most expensive items on the third level, and in four games there was a winning solution that involved pairing the third and fifth most expensive items on the third level. In one game, pairing the fourth and fifth most expensive items on the third level would have led to victory.

Key Tip

If able to identify the most expensive, second most expensive, and least expensive items with some degree of certainty, play for $100,000 by placing the most expensive item on the top level and the second most expensive and least expensive items on the third level. Place the third most expensive item on the bottom level and the fourth and fifth most expensive items on the second level. Otherwise, order in ascending price on the first three levels and stop at $10,000.

PLINKO

Plinko, illustrated in Figure 15.5, is a perennial contestant favorite on *The Price Is Right*. Numerous contestants show up with *Plinko* affinity attire, hoping to play *Plinko*, and for good reason. The $12,200 median win value is impressive, as is the $7,050 in median earnings. It was the second highest earnings cash game behind *Pay the Rent*, and the possibility of one's chip landing in the $10,000 slot is thrilling.

Contestants are generally effective in getting additional *Plinko* chips beyond the one chip they start with, averaging an additional 2.9 out of 4 possible *Plinko* chips, for a 73% success rate. Thirteen out of forty-seven times they earned all four additional *Plinko* chips, while twenty times they earned three more chips, twelve times two more chips, and two times one more chip. The binomial distribution with 73% success rate would suggest 0/1/2/3/4 additional chips with probabilities 1%/6%/23%/42%/28%, similar to what was observed.

On the *Plinko* board, one might be tempted to overthink placement of the chip. If the chip has an equal chance of bouncing right or left off the pins, the ideal release point is dead center. True, you risk falling into the zero slots, but two or more zeros for one $10,000 slot is not a bad trade-off. Still, the $10,000 slot is very hard to land. Only 26 of 185 *Plinko* chips fell in the

Figure 15.5 *Plinko. Illustration courtesy of Omar Faruk*

$10,000 slot, once per game in twenty-two games and twice per game in two games. Altogether, the chips fell as shown in Figure 15.6. This includes special games in which the slots had higher prize values. Here I map the higher values into regular *Plinko* board values for descriptive purposes. The average chip earned the contestant $1,800 applying regular slot values.

Landing Slot	$10,000	$0	$1,000	$500	$100
Drops (out of 185)	26	51	51	43	14
Percentage of Drops	14.1%	27.6%	27.6%	23.2%	7.6%

Figure 15.6

Many studies have calculated the odds of landing in the $10,000 slot depending on where the chip is dropped and assuming a random walk with side pins. Figure 15.7 shows these probabilities, with the starting columns (i.e., drop column) shown on the left, and the slot values shown at the top. Start columns A–H reflect the nine openings on the *Plinko* board where a contestant can drop the chip, from left to right. Before the *Plinko* chip reaches the bottom, it descends twelve rows (including the starting row) and can bounce off any of five side pins on either side of the *Plinko* board.

Start column	$100	$500	$1000	$0	$10000	$0	$1000	$500	$100
A	.226	.387	.242	.107	.032	.006	.000	0	0
B	.193	.346	.247	.137	.057	.016	.003	.000	0
C	.121	.247	.241	.196	.121	.054	.016	.003	.000
D	.054	.137	.196	.226	.193	.121	.054	.016	.003
E	.016	.057	.121	.193	.226	.193	.121	.057	.016
F	.003	.016	.054	.121	.193	.226	.196	.137	.054
G	.000	.003	.016	.054	.121	.196	.241	.247	.121
H	0	.000	.003	.016	.057	.137	.247	.346	.193
I	0	0	.000	.006	.032	.107	.242	.387	.226

Figure 15.7 *Plinko* Slot Probabilities Based on Drop Position (Rows Add to 100%)

Per Figure 15.7, the probability of landing in the $10,000 slot is 22.6% if the chip is placed directly in the center. The probability falls to 19.3% if the chip is dropped one position to the right of center or one position to the left of center. It falls further to 12.1% if dropped two positions to the right of center or two positions to the left of center, and lower if dropped even further from the center.[‡]

I did not observe the success of landing in the $10,000 slot as predicted by Figure 15.7. This is due in part to contestants placing the chip in the middle slot only 18% of the time, as seen in Figure 15.8.[§] But dropping from either the middle column or from one column to the left or right of the middle column also led to lower success than the probabilities would suggest, albeit with a limited data set. Why is this the case?

[‡]Robert Burks and Michael Jaye, "The Price is Right Again," *Journal of Statistics Education* 20, no. 2 (2012).

[§]Data set captures only 111 of the 185 chips dropped during seasons 47 and 48, as CBS All Access had removed seasons 47 and 48 by the time I ran this analysis, limiting my data set to *Plinko* games available through videos posted online.

Column	A	B	C	D	E	F	G	H	I	Total
# Drops	0	3	15	22	20	26	20	4	1	111
Share of Drops	0%	3%	14%	20%	18%	23%	18%	4%	1%	100%
Lands $10,000 Slot	0	0	2	3	3	5	3	0	0	16
Observed Hit Rate		0.0%	13.3%	13.6%	14.4%	19.2%	15.0%	0.0%	0.0%	14.4%
Predicted Hit Rate*	3.2%	5.7%	12.1%	19.3%	22.6%	19.3%	12.1%	5.7%	3.2%	16.6%

* Predicted Rate per Figure 7 Probabilities

Figure 15.8 Chip Placement and $10,000 Slot Success

Players occasionally disadvantaged themselves by dropping the chip from above the board instead of placing it in the top row. This led to chip velocity and perhaps some chip direction before the chip struck the first pin, making the first pin hit—and subsequent pin hits—not a random walk. So instead of the chip having a 50/50 chance going left or right upon hitting the first pin, it might have had a 60% chance of going one direction or the other. Four out of twenty drops from the middle column (i.e., column E) were dropped a meaningful distance above the starting row, and four were dropped a modest distance above the starting row. One of these eight drops fell in the $10,000 slot, but three fell in the $100 slot, which is supposed to be only 3.2% likely for a "random walk" drop from the center slot. And three of these eight drops hit the side pins, also very unlikely in a "random walk" scenario.

> **Key Tip**
>
> Place the *Plinko* chip in the center slot and release it from the top row, taking care not to release it from above the board.

PUNCH-A-BUNCH

Punch-a-Bunch seems like a great cash game with the opportunity to win up to $25,000, but the distribution of cash payouts is much more conducive to smaller wins.

On the standard *Punch-a-Bunch* board, there are fifty possible punches (5 × 10), with the following values: one worth $25,000, two worth $10,000, four worth $5,000, eight worth $2,500, ten worth $1,000, ten worth $500, ten worth $250, and five worth $100. There is a total of $103,000 on the board, for an expected value on any one punch of $2,060. Median earnings of $2,400, ranking seventy-third out of seventy-four pricing games, should come as no surprise.

Note that a contestant with one punch remaining should stop with a prize of $2,500 or higher, as $2,500 is above the $(\$103,000 - \$2,500)/49 = \$2,050$

expected value of the cash prizes remaining on the board. In evaluating multiple punch scenarios, the contestant should stop play if he draws a card greater than the expected value of continued play. With multiple punches, the expected win value in *Punch-a-Bunch* increases from $2,060 to as high as $4,832 as the contestant gets more opportunities to draw a higher cash card.

For interested readers, the appendix with Supporting Math offers a detailed mathematical discussion of how to calculate the expected win value and desired stopping point in *Punch-a-Bunch*. *Notably, the expected win value is such that contestants should stop at $2,500 or higher with one or two total punches or with three or four punches when only one punch remains. With two or more punches remaining, the contestant should stop only with a cash card of $5,000 or higher.*

Figure 15.9 shows the average win value during seasons 47 and 48 relative to the number of total punches and versus the expected win value when following the strategy articulated here. For each number of punches, observed performance trailed expected performance, leading to a paltry $2,400 in median earnings. The $2,400, however, reflects more bad luck than bad play. Across ninety-three revealed punches, there were no $25,000 punches and only one $10,000 punch, with an average punch of $1,550. A $10,000 or $25,000 card—or lack thereof—can throw off the averages meaningfully.

Punches	1	2	3	4
Avg Observed Win	$ 1,000	$ 1,963	$ 2,778	$ 2,180
Avg Predicted Win	$ 2,060	$ 3,160	$ 4,055	$ 4,832

Figure 15.9 *Punch-a-Bunch*: Observed versus Predicted Win Value

Contestants made limited mistakes in terms of stopping versus playing for more money. There were four probabilistic errors, but they did not affect total winnings in *Punch-a-Bunch*. I highlight four close calls and whether the outcome was helped or hurt by the contestant's decision.

- January 14, 2019: The contestant stops at $5,000 on his second punch. He would have ended up with $10,000 on his fourth punch (*right call*, despite missing out on $10,000).
- March 1, 2019: The contestant turns away $2,500 on his third punch, only to end up with $250 on his fourth and final punch (*wrong call*, negative outcome).
- April 9, 2019: The contestant stops at $2,500 on his second of three punches (*right call*, positive outcome, as the third punch would have only amounted to $250).
- October 15, 2019: The contestant plays on after $2,500 on second of three punches. Wins $10,000 on third pick (*wrong call*, positive outcome).

Lack of success was not for lack of winning punches. On average, contestants won 3.3 punches, with fifteen of twenty-nine contestants earning four punches, nine contestants winning three punches, four contestants winning two punches, and one contestant winning one punch. To win small items, bear in mind that when the displayed price was $100 or more, the correct pick was lower thirty-six of those thirty-eight times. Conversely, when the displayed price was $50 or less, the correct price was higher fifty-two of fifty-six times. Overall, about half of the small items were higher than the displayed price and about half were lower.

Key Tip

With one or two punches, stop at $2,500 or higher. With three or four punches *and* two or more punches remaining, stop at $5,000 or higher. To win punches, go lower if shown a price of $100 or more.

TIME IS MONEY

One of the more chaotic games on *TPIR*, *Time Is Money* tests contestants' speed and composure under pressure. Across twenty-eight games, contestants won nine times with a median prize value of $18,900. Most wins were near the full $20,000. Although only two contestants won the full $20,000 on their first try (within the allotted ten seconds), three contestants quickly rearranged items as the "cash clock" started winding down and won about $19,000. Four other players won between $12,000 and $16,000, and three "non-winning" contestants came away with sub-$3,000 in winnings just before time ran out. The other sixteen contestants came up empty-handed.

In terms of possible combinations, there are three pricing options for each grocery item, creating $3 \times 3 \times 3 \times 3 \times 3 = 243$ possibilities. However, in practice, there were never more than three grocery items on each pricing table, and in all but three instances of the game, the five prizes were split 2/2/1 across the three tables (not necessarily in that order). Having three items on a table was a more difficult setup, and contestants came up empty-handed in all three of those games.

In a 2/2/1 scenario, there are 90 feasible combinations out of 243, as one can remove the following 153 combinations:

- Three combinations with all five prizes on one table (five on each of three tables);
- Thirty combinations with four of five prizes on one table (five pairings of four on any of three tables and two possibilities for the fifth item: $5 \times 3 \times 2 = 30$ pairings); and

- One hundred and twenty combinations with three of five prizes on one table (ten pairings of three items on any of three tables and four possibilities for the remaining two items: $10 \times 3 \times 4 = 120$ pairings)

Sticking with the 2/2/1 scenario, if the contestant has certainty about the price of one item, the ninety remaining possibilities are effectively cut in three to thirty, as the known item can no longer be on any of the three tables. If the contestant has certainty about the price of two items, the number of possible options is further cut to twelve or six, depending on whether the second known item is in a different price bucket (twelve options) or the same price bucket (six options) as the first known item.** Note that if the second known item is in the same price bucket as the first known item one-third of the time, then on average we still can eliminate twenty out of thirty, or two-thirds of the possibilities: $\left(18 \times \frac{1}{3}\right) + \left(18 \times \frac{1}{3}\right) + \left(24 \times \frac{1}{3}\right) = 20$ eliminated possibilities.

Knowing the price bucket for two items is a fair baseline for contestants starting the game, as one item is usually identifiable as inexpensive (i.e., $0–$2.99 bucket) and one as expensive (i.e., $6 and more bucket), reducing the game to twelve options. There is ample time to try many of the potential twelve possibilities, provided the contestant is deliberate and fast. Ideally, he would try what he believes is the most likely combination in the first ten seconds and then try progressively less likely combinations during the countdown phase. If he has conviction that one of the other three items is not in the $0–$2.99 bucket, for example, that would put three of the twelve choices in the less likely camp. And if he has near certainty on a different third item, that would prioritize two, three, or four of the remaining nine choices.††

Consider the episode from June 28, 2019. The contestant had to place a box of twenty tea bags, a pack of three Brillo pads, 1.125-ounce shoe polish, a ten-ounce bag of mini marshmallows, and Gold Bond cream (13 ounces). The Brillo pads were very likely $0–$2.99, and the Gold Bond very likely $6 or more, whereas the marshmallows and tea bags were very likely not $6 or more, and the shoe polish very likely not $0–$2.99. If those inferences are all correct and the answer is a 2/2/1 combination, there are only five possible combinations, as shown in Figure 15.10.

**When the second known item is in a different price bucket, we cannot eliminate two-thirds of the choices, as any of the remaining prizes can pair with either of the known items in a 2/2/1 setup. When the second known item is in the same price bucket, we can eliminate more than two-thirds of the choices because nothing else can be put in the price bucket with the two known items. It is a "known" bucket.

††Near certainty on the $0–$2.99 bucket would prioritize three of the remaining nine choices. Near certainty on the $3.00–$5.99 bucket would prioritize four choices, whereas near certainty on the $6 and more bucket would prioritize two choices.

Choice	0 - $2.99	$3 - $5.99	$6+
1	Brillo Pads, Tea Bags	Marshmallows	Gold Bond, Shoe Polish
2	Brillo Pads, Marshmallows	Tea Bags	Gold Bond, Shoe Polish
3	Brillo Pads, Tea Bags	Marshmallows, Shoe Polish	Gold Bond
4	Brillo Pads, Marshmallows	Tea Bags, Shoe Polish	Gold Bond
5	Brillo Pad	Tea Bags, Marshmallows	Gold Bond, Shoe Polish
Correct	**Brillo Pads, Marshmallows**	**Tea Bags**	**Gold Bond, Shoe Polish**

Figure 15.10

If I were playing, I would have viewed the shoe polish as likely $6 or more, and I would have expected either the tea bags or marshmallows to be priced less than $3. Consequently, I would have prioritized options one and two, winning on my initial or second try. The contestant here was a bit lost. On his first choice he had the mini marshmallows, tea bags, and Gold Bond cream correctly placed, with the shoe polish on the $0–$2.99 table and the Brillo pads on the $3–$5.99 table. Despite having three picks initially correct, the contestant failed to get the prizes correctly placed before the cash clock ticked down to zero. He never placed the shoe polish on the expensive table, which was his downfall.

Contestants can get frazzled with incorrect choices when playing against the clock, sometimes removing the "certain" items from the appropriately priced table. However, the majority of losses (eleven out of nineteen) followed initial guesses with zero, one, or two items correct. Contestants never earned any money when they failed to get three items correct on their initial guess. Conversely, contestants with four items correct on their first try won five of seven times, and contestants with three items correct on their first try won two of eight times.

Key Tip

Time Is Money is very difficult. Quickly decide if you can assign certain items to certain price buckets or rule out certain items for certain price buckets. Work fast to try the remaining options.

16

Grocery Games (Non-Cash)

Grocery games are a staple of *The Price Is Right* (*TPIR*), no pun intended. There are eleven grocery games, four that are played for cash (covered in chapter 15) and seven that are typically played for non-cash prizes, covered in this chapter. Non-cash grocery games were played 163 times during seasons 47 and 48, or almost once every two episodes. Only once were two grocery games played in the same episode.

Non-cash grocery games can be some of the easiest and some of the hardest pricing games, as seen in Figure 16.1. On one end of the spectrum is *Bullseye*, which was won 91% of the time, and *Now or Then*, won 88% of the time. On the other end was *Check Out*, which was won 26% of the time. Category-wide, the 57% win rate was in line for other categories of non-car pricing games. But median prize values tend to run in the $9,000 to $11,000 range, higher than the $7,000 to $9,000 range for two item, three item, four item, and price guess games. The effect is a relatively high $5,400 level of median earnings.

Game	Played	Wins	Losses	Win Rate	Average Prize	Median Prize	Avg Win	Median Win	Min Win	Max Win	Median Earnings	Earnings Rank
Bullseye	23	21	2	91%	$11,494	$12,222	$11,552	$12,222	$5,000	$25,000	$11,159	3
Check Out	23	6	17	26%	$10,937	$10,000	$12,481	$11,532	$5,018	$19,575	$3,008	67
Grocery Game	17	7	10	41%	$11,807	$10,863	$10,136	$8,350	$5,222	$20,940	$4,473	50
Hi Lo	21	11	10	52%	$10,164	$8,900	$8,927	$8,700	$5,360	$17,530	$4,557	45
Now or Then	17	15	2	88%	$8,920	$6,732	$8,423	$6,732	$5,000	$14,846	$7,432	17
Pick-A-Pair	29	19	10	66%	$8,764	$8,558	$8,893	$7,800	$5,046	$16,675	$5,110	40
Vend-O-Price	33	14	19	42%	$9,786	$8,753	$9,705	$7,721	$5,438	$26,340	$3,275	63
Total/Wtd Avg	163	93	70	57%	$10,177	$9,412	$9,869	$9,003			$5,400	

Figure 16.1 Statistical Summary—Grocery Games

- *Bullseye*: A contestant is shown five grocery items, one of which has a hidden bullseye designation on the back. He has three chances to pick an item in a certain quantity, the total cost (i.e., price × quantity) of

165

which he believes will fall in the $10 to $12 range. If he is correct on his first try, he wins. Otherwise, he tries again with a different item, and if incorrect, gets a third try with a third item. If a guess totals between $2 and $10, that grocery item lands on the target board. Should none of the player's three guesses land in the $10 to $12 range, he can still win if one of his grocery items on the board has the bullseye behind it. A guess that exceeds $12 takes that grocery item out of play for the bullseye.

- *Check Out*: The contestant is shown five grocery items. He guesses a price for each. To win, the total guessed price of the five items must fall within $2 of the actual total price of the five items. The contestant need not be close on the price of any individual item to win.
- *Grocery Game*: The contestant is shown five grocery items and must buy one or more items in a specified quantity such that the total price amounts to between $20 and $22. The game ends when the contestant wins, goes over $22, or runs out of grocery items to buy.
- *Hi Lo*: The contestant is shown six grocery items and must pick the three most expensive items of the six. If correct, he wins.
- *Now or Then*: The contestant is shown six grocery items and associated prices placed in six wedges of a circle. Some prices are today's price and some a historical price, the historical price falling in a specified month, ranging from May 1998 to July 2010. To win, the contestant must correctly identify three adjacent items in the circle as "now" or "then." The game continues until the contestant gets three in a row or has no further possibilities to identify three in a row.
- *Pick a Pair*: The contestant is shown six grocery items consisting of three pairs of equally priced items. The contestant tries to pick one equally priced pair of items. If he is incorrect on his first try, he gets a second chance to keep one of the two selected items and try to find a correct price match from one of the three remaining items. After the second try, win or lose, the game is over.
- *Vend-O-Price*: A vending machine reveals three shelves, each with a grocery item in a given quantity. The contestant must select the most expensive of the three shelves (price × quantity) to win.

BULLSEYE

With a 91% win rate (21 out of 23), *Bullseye*'s $11,150 median earnings were the highest of any non-car pricing game. This reflected the high $12,200 median prize value, the ease of the game, and good luck.

Why is the win rate in *Bullseye*, illustrated in Figure 16.2, so high? First, the contestant gets three tries to land between $10 and $12 using a specified quantity of a specific grocery item. Second, there are no "non-winning" items—a certain quantity of each item will total between $10 and $12. Prices ranged between $1.49 on the low end and $10.99 on the high end, although only two of forty-eight revealed items were priced above $6.00. Third, the bullseye will match one of the five items.

If a player had used the simplistic strategy of choosing two of each item, he would have had three items on the target board in all but the two games in

Figure 16.2 *Bullseye. Illustration Courtesy of Omar Faruk*

which an item was priced above $6. With 21% of selected items priced in the $5 to $6 range during seasons 47 and 48 (10 out of 48), 51% of games would have ended with a win in the $10 to $12 range.* Otherwise, the player would have placed three items on the board in almost every game, meaning a nearly 60% chance (3 of 5) of winning with the bullseye. We assume a one in five chance (20%) of an item being in the $5 to $6 range and a one in ten chance (10%) of an item being in the $10 to $12 range and thus not landing on the board when picked in twos. The win rate using the two-of-each-item strategy would be about $48.8\% + (54\% \times 51.2\%) = 76.4\%$.

Instead of picking two of each item, a player who believes he has at least a 30% chance of landing in the $10 to $12 range should target that range on each pick for at least a 76% chance of winning. Consider the following:

1. A 30% chance per guess allows him to win in the $10 to $12 range with probability $1 - 0.7 \times 0.7 \times 0.7 = 0.657$, or 65.7%.
2. Assume the player goes over and falls short with equal frequency when he fails to land in the $10 to $12 range, similar to what is observed. Then each try will have a 35% chance of landing on the board.
3. The binomial distribution indicates a 1/8 chance of zero non-winning tries landing in the $2 to $10 range, a 1/8 chance of all three tries landing in the $2 to $10 range, and a 3/8 chance of one or two tries landing in the $2 to $10 range.
4. If the player does not win with a $10 to $12 strike, his chance of winning sums up the bullseye probabilities across scenarios with three, two, and one try on the board: $\left(\frac{1}{8} \times \frac{3}{5}\right) + \left(\frac{3}{8} \times \frac{2}{5}\right) + \left(\frac{3}{8} \times \frac{1}{5}\right) + \left(\frac{1}{8} \times \frac{0}{5}\right) = \frac{3}{40} + \frac{6}{40} + \frac{3}{40} + \frac{0}{40} = \frac{12}{40} = 0.3$, or 30%.
5. The probability of winning *Bullseye* under this construct is $65.7\% + (1 - 65.7\%) \times 0.3 = 76.0\%$.

During seasons 47 and 48, players won by landing in the $10 to $12 range 33% of the time (16 out of 48 tries), so a 30% success rate of landing in the $10 to $12 range matches observed performance.

Below I show how the odds of winning *Bullseye* vary with the chance p of getting an item in the $10 to $12 range, assuming items that do not fall in the $10 to $12 range are as likely to land on the board as they are to exceed $12. The formula is $P(\text{win}) = 1 - (1 - p)^3 + 0.3 \times \left(1 - \left(1 - (1 - p)^3\right)\right) = 1 - (1 - p)^3 + 0.3 \times (1 - p)^3$.

---

*Binomial distribution with $p = 0.21$, $n = 3$, $k = 1, 2$, or 3 yields 50.7%. Using $p = 0.20$ yields 48.8%.

P (land on $10–$12, per try)	20%	30%	40%	50%	60%
Total Win Probability	64%	76%	85%	91%	96%

I believe contestants can land in the $10 to $12 range 35 to 40% of the time and win 80 to 85% of the time if they avoid egregiously expensive guesses. That win rate is better than the pick-two-of-each-item strategy, as most wins result from landing in the $10 to $12 band. Nor should contestants guess too conservatively, lest they reduce their chances of landing in the $10 to $12 range, even if this means some guesses not landing on the board. Contestants seemed to understand this, with sixteen out of forty-eight tries landing in the $10 to $12 range, seventeen tries being too high, and fifteen tries being too low.

Bullseye performance also benefited from chance. There were five out of seven bullseye wins when the player failed to land in the $10 to $12 range, despite his having only two items on the board in four of the seven games and one item on the board in the other three games. Chance would have suggested only 2.2 bullseye wins in seven games given the eleven items placed on the board.

Key Tip

Go for the $10 to $12 range. Don't purposely err low, but avoid large overbids.

CHECK OUT: SEEING THE BIG PICTURE

In contrast to *Bullseye*, *Check Out* was one of the hardest non-car games, ranking sixty-seventh in median earnings, with six wins in twenty-three tries. If not for the more generous $10,000 median prize value, the game would have stacked up even more poorly.

Although *Check Out* is a difficult game, contestants also execute poorly, failing to see the big picture. Most contestants choose the price of an item based on their best guess for that item rather than targeting a total aggregate price. In practice, there should be little difference between totaling one's best guess for five grocery items and guessing the total price of five grocery items. But sometimes one repeatedly errs low or high, and the errors do not offset. During seasons 47 and 48, there was a fairly narrow range for the aggregate sum.

Each contestant should guess the price of the first four items "straight up" and then guess the price of the fifth item (always the most expensive) with a desired aggregate bid in mind. Not once during seasons 47 and 48 did the contestant's total go over by more than the price of the last item—the highest "total overbid" was $4.25 versus a fifth item valued at $6.27. Therefore, the contestant could have always guessed low enough on the fifth item to achieve his desired aggregate total (i.e., his first four guesses did not invalidate this strategy).

The sum total of the five grocery items ranged from $14.31 on the low end to $23.13 on the high end, with an average of $19.29 and a median of $20.55. Had the contestant always chosen the total price to be $19.50, creating a $17.50 to $21.50 range, he would have won fourteen out of twenty-three iterations of the game, losing the five times that the total price was above $21.50 and the four times that the total price was below $17.50. To reach a total of $19.50, the contestant could have even guessed zero for each of the first four items and $19.50 for the fifth item.

Interestingly, more than half of the variation in price in *Check Out* related to the last, most expensive item. Consider the interquartile range (IQR), the difference between the first quartile and the third quartile. The sum of the first four items had an IQR of $10.07 to $12.26 with a median of $11.50, whereas the fifth item had an IQR of $6.49 to $9.64, with a median of $8.99.

With this knowledge, it is not hard to devise a strategy. Start with a total price of $20.[†] If the items in aggregate seem materially more expensive than $20, which usually coincides with the last item costing more than $8, raise the total to $21. If the items in aggregate seem materially less expensive than $20, which usually coincides with the last item costing less than $7, lower the total to $18. Or, alternatively, choose prices for the first four products totaling $11.50 and guess the price of the last product. The last item had an average price of $8, a median price of $9, and three times maxed out at $9.99. In three games, however, the last item was priced very low at $4.99, leading to low aggregate totals of $14.55, $16.34, and $14.31, respectively.

Choosing a $20 total every time would have produced thirteen wins in twenty-three games during seasons 47 and 48, and flexibly adjusting upward to $21 or downward to $18 would have led to seventeen wins. I believe the $20 guess should lead to at least a 50% win rate in future seasons and even better with adjustments.

Choosing a price for each item without regard to the total can backfire, as occurred on September 26, 2018. For items 3–5, the contestant's total guess of $15.50 was close to the $15.97 actual. But being far too high on the first two items ($5.98 guess versus $3.09 actual) negated that achievement, causing the loss. Notably, three of the six wins involved the contestant reversing a running tally that was too low or too high with his guess on the fifth item. We do not know if this was intentional or serendipitous. Most losses, however, involved a bid on the last item that exacerbated an insufficient or excessive subtotal.

[†]The contestant could default to $19.50, but $20 is simpler and can accommodate some future inflation.

Key Tip

Choose prices for the first four items that add to $11.50. Then price the fifth item. The average/median price of the fifth item was $8/$9 in seasons 47 and 48, but in some cases the fifth item was priced as low as $5.

Seeing the Big Picture

In merger arbitrage investing, standard practice is to buy the stock of the target after a deal has been announced and hold the target until the deal is completed. Although the target's stock typically moves up sharply on an announced bid, the probability of the deal closing is usually high, and there is a nice spread that is mostly uncorrelated with the market that closes between the time the deal is announced and when the deal closes. Individuals focused on merger arbitrage can usually avoid failed deals. But there is also a bigger picture to contemplate. Sometimes the buyer has made a bid to ward off a third party from bidding on its own business. An example is Hillshire bidding to acquire Pinnacle Foods in 2014. JBS had made overtures to acquire Hillshire prior to Hillshire's bid for Pinnacle Foods. Shortly after the Hillshire bid for Pinnacle and well before the Pinnacle deal had closed, JBS made a compelling bid for Hillshire. Pinnacle's stock gave back most of its gains tied to Hillshire's initial bid. A merger arbitrage investor who acquired Pinnacle's stock after Hillshire's announced bid would have been meaningfully in the red on its investment. But someone who recognized Hillshire also to be in play could have come out way ahead, especially as Tyson Foods ended up topping JBS's bid for Hillshire and ultimately acquiring it.

GROCERY GAME: THE GRADUAL APPROACH

Played only seventeen times and won only seven times, *Grocery Game* is also much easier with strategic play. Three of ten losses involved going over $20 to $22 on one's the first guess, a fatal mistake given the option of building gradually to the $20 to $22 range. Four of the other seven losses involved being over the $22 by less than $0.20, a dose of bad luck.

The contestant should build to the $20 to $22 range *gradually* to avoid going over. To do so, he needs less expensive items available as he approaches the $20 to $22 range. On the January 2, 2019, show, the contestant's tally was $19.93 after selecting three $3.69 fabric softeners, two $2.99 paper towels, one $1.59 toothpaste, and one $1.29 water. He then had no option but to take one 3.5-ounce icing gel, priced at $3.99, and he ended up at $23.92, above the $20 to $22 range. Although he proceeded slowly and deliberately, had he chosen one icing gel third, he would have won with $21.04. Or he could have instead selected two waters after the toothpaste, when he had a $18.64 subtotal. It is unlikely the bottled water would have cost more than $1.68 a bottle, putting his tally over $22.

Going for broke on one item is rarely a sound strategy. Some items do not total between $20 and $22 at any quantity, including eleven of the seventeen items priced at $3.69 or higher in seasons 47 and 48. Moreover, estimation errors are magnified in larger quantities. Better to get close with an expensive item or two and then move toward the target in a measured fashion with less expensive items. Notably, one and sometimes two items are priced less than $2. These grocery items allow for a win at any point in the game. Contestants won four out of eight times (50%) with three to five selections and only three out of nine times (33%) with one or two selections, despite the aforementioned loss with the icing gel when it was the only item remaining.

Key Tip

Start with an expensive item to land in the $15 to $20 range. Use a mid-priced item to get near $20 or just above $20 for the win. If needed, use a sub-$2 item to top up to the $20 to $22 range.

The Gradual Approach

Real options are a core part of business strategy. Stanley Black & Decker (SWK), a maker of tools and industrial products, demonstrated the benefits of the gradual approach with its acquisition of MTD Products. In September 2018 SWK agreed to acquire 20% of the outdoor power equipment company for $234 million. The price for the equity stake was low relative to MTD's $2 billion in sales because the price was a multiple of MTD's profit, and MTD's operating margin was only 4%. Importantly, the deal included an option for SWK to buy the remaining 80% of MTD starting in July 2021. SWK also got a board seat and formed a team to improve the profitability of MTD's operations, which allowed it to nearly double MTD's operating margins. SWK exercised its option to acquire the remaining 80% of MTD in late 2021, confident that it could further improve MTD's margins. SWK paid a multiple tied to the average profitability at the time of the initial 20% purchase (i.e., low margin) and the time of the 80% option exercise (i.e., better margin). SWK thus acquired the remaining 80% of MTD at a good valuation, while benefitting from the deurbanization trend brought on by COVID, at least until interest rates spiked up. By proceeding gradually, SWK preserved its optionality, rather than pursuing an all-or-nothing approach that might have backfired.

HI LO

Hi Lo had middle-of-the-road performance. A 52% win rate (11 out of 21) and $8,900 median prize value drove $4,550 median earnings, ranking forty-fifth. Usually, two of the items are materially more expensive than the bottom three items and the third most expensive item is close in price to the third least expensive item, with an average $1.55 difference. In eight of the ten losses the contestant picked the third least expensive prize in his "Hi 3" prizes, whereas the other two losses involved him picking two of the three least expensive items in his "Hi 3."

    Hi Lo is very similar to *Grand Game*. In *Hi Lo* the contestant selects the three most expensive items rather than the four most expensive items in *Grand Game*. The odds of success should be slightly lower in *Hi Lo* for an identical set of groceries, making the 52% win rate in *Hi Lo* versus 30% in *Grand Game* surprising. Of course, the groceries are neither identical nor equally difficult. Similar to *Grand Game*, *Hi Lo* losses were mostly caused by confusing prices, such as $10-plus trash bags, although better luck and fewer outlier prices led to better win rates in *Hi Lo*.

NOW OR THEN

Now or Then was one of the easiest games on *TPIR*, with an 88% win rate (15 out of 17) and median earnings of $7,450. Median earnings for *Now or Then* are based on average win value, as the median win value is skewed downward by many lesser (i.e., sub-$7,000) prizes.

    To understand the high win rate in *Now or Then*, consider the probability of getting three consecutive heads in a "wrap around" sequence of six coin tosses, assuming you call each toss heads. There are sixty-four possible combinations of heads and tails in a set of six tosses, namely 2^6. There are twenty combinations among those sixty-four that yield three heads in a row in a normal sequence, and five more combinations that wrap around. The probability of winning the coin toss game is thus $25/64 = 39.1\%$.

    This coin toss game is the logical equivalent of getting three adjacent tries correct in *Now or Then* if the odds on each try are 50%. It does not matter whether you call for three heads, three "nows," or three "thens," because each coin toss and grocery is independent and has a consistent 50% probability. During seasons 47 and 48, contestants did much better than random chance, winning 77% of tries (50 out of 65). One can mathematically derive the odds of victory as $6p^3 - 6p^4 + p^6$ if p is the probability of success on each try. For p = 77%, this formula predicts an 84% chance of success, similar to the 87% observed.[‡]

[‡]Please visit www.popculturemath.com to see the derivation of the odds of victory in *Now or Then*.

How did the fifteen wins play out? Eight wins resulted from three consecutive correct picks, with five wins on the fourth try after missing the third try, and two wins on tries four, five, and six after erring on the second and third tries. One of the wins with six tries, on June 7, 2019, is captured in Figure 16.3. The contestant got "now" correct for spaghetti, then erroneously called the

Figure 16.3 *Now or Then. Illustration Courtesy of Omar Faruk*

razor "then" and the baking spray "now." He went on to correctly identify the $4.79 syrup "now," the $2.99 wipes "then," and the $9.99 lotion "now."

"Now" products were much more common than "then" products. Of the sixty-five grocery picks, forty-six were "now" priced and nineteen were "then" priced. This sizable skew toward "now" would be expected to occur about 0.1% of the time if "now" or "then" were equally likely. Contestants guessed "now" forty-four out of sixty-five times and were correct with thirty-eight out of forty-four "now" picks, versus twelve out of twenty-one "then" picks. I suspect that there are two "then" groceries per board, although not necessarily directly across from one another, creating the opportunity for the contestant to win with three "nows" in a row.

Interestingly, *if there are* always *two "then" products on the board, a contestant can design a strategy to guarantee victory 100% of the time.* I encourage folks to review the 2013 *Slate* article on *The Price Is Right* by Ben Blatt for more detail. Essentially one guesses "now" for each of the three groceries on one half of the board, then conditionally guesses "now" or "then" in a strategic order for the remaining three grocery items.

Key Tip

There are likely two "then" choices per board. Try and identify one "then," start with that product, and then find adjacent "now" products to complete three in a row.

PICK A PAIR

Contestants did well at *Pick a Pair* and potentially could have done a bit better. With a 66% win rate (19 out of 29) and an $8,550 median prize value, *Pick a Pair* generated $5,100 in median earnings, ranking fortieth.

Of the nineteen wins in *Pick a Pair*, ten were on the first try and nine were on the second try. The player got the first pair correct 34% of the time (10 out of 29), which is modestly better than a one in five chance if randomly picking from the other five grocery items. It is also similar odds to ruling out two choices and picking at random from the three remaining choices. Similarly, nine wins and ten losses on the second attempt equates to a 47% win rate versus a one in four random guess. Performance relative to random chance was similar on one's first try (1.7 × random chance) and second try (1.9 × random chance).

If *Pick a Pair* were random, the contestant would have a one in five chance of winning on his first try and a one in four chance of winning on his second try, which occurs the four in five times he fails to win on his first try: $P(\text{random win}) = \frac{1}{5} + \left(\frac{4}{5} \times \frac{1}{4}\right) = \frac{2}{5}$, or 40%.

The 66% observed win rate is meaningfully better than random chance. The contestant likely has a good idea as to one prize in the low- or high-priced category, even if he does not know which prize to pair it with. The contestant should focus on pairing one of these low- or high-priced prizes. Although he may not get the initial pairing correct, he can usually tell based on the prices revealed whether each item falls in the low, middle, or high price category.

For the second pairing, the contestant should choose to keep an item in the low or high price category. He can then eliminate choices at the opposite end of the price spectrum, likely setting up a two-choice guess with 50/50 odds or better. In contrast, if he keeps a mid-priced item, process of elimination becomes less effective, as it is harder to exclude items from the mid-priced bucket.

On twelve occasions the contestant did not get the first pairing correct. Five or six times he went on to keep the item for his second pairing that likely had the middle price, allowing him to eliminate fewer choices for his second pairing. In the game on June 19, 2019, the contestant failed in his first pairing of $8.99 pain relief and $5.49 baby food. But he kept the baby food, which likely had the middle price, instead of the pain relief, which certainly had the high price. He then paired the instant coffee with the baby food, but the instant coffee was $8.99. Had he kept the pain relief, he would have instead chosen between instant coffee or cookies, since the wipes and paper plates clearly did not cost $8.99. The other $5.49 item, in contrast, could have been three of the four remaining items: baby food, wipes, or cookies.

Keeping a low- or high-priced item on the second pairing should allow for a 50 to 60% win rate on the second pairing after a one in three win rate on the first pairing, for a 67 to 73% chance of winning *Pick a Pair*.

Key Tip

Start with a likely high- or low-priced item and try to pair it. If incorrect in the initial pairing, keep an item that is high or low priced and try to pair up that item. Pairing a high- or low-priced item on one's second try should aid the process of eliminating other choices.

VEND-O-PRICE

Vend-O-Price was the most played non-cash grocery game in seasons 47 and 48. It was won 42% of the time (14 out of 33), with a $8,750 median prize value and $3,300 in median earnings.

The lower shelf (i.e., the most expensive item in the lowest quantity) was the correct pick twelve times, the middle shelf (i.e., the mid-priced item with the middle quantity) thirteen times, and the upper shelf (i.e., least expensive item with the highest quantity) eight times. Players chose the lower shelf five times, winning once; the middle shelf twenty-four times, winning eleven times; and the upper shelf four times, winning twice. It is unclear if the middle shelf showed up more frequently due to chance or to make the game easier.

The number of items on each shelf was relatively consistent. All but two times the lower shelf had three or four items; all but once the middle shelf had five or six items, while the top shelf had seven to ten items. When there were an extreme number of items on the upper and lower shelf, they were never the right choice. The top shelf was not the correct pick the seven times there were ten items on the top shelf, and the bottom shelf was not the correct pick the two times there were two items on the bottom shelf.

Not surprisingly, players struggled with smaller price differentials, winning only two out of nine times when there was a sub-$4 differential between the most expensive shelf and the second most expensive shelf. They performed very well when the price differential was large, winning five out of seven times with a $7-plus differential.

Key Tip

If there are ten items on the top shelf and two items on the bottom shelf, avoid those shelves. The top shelf may be a bit less likely to be the correct pick. Otherwise, use your best judgment.

17

Small Item Pricing Games

Small item pricing games are like grocery games, except here the contestant is pricing items typically worth less than $200 to win the game's prize. These games have favorable setups, as seen in Figure 17.1. Three of the five highest-ranked median-earnings games not played for cars were small item pricing games. In *Bonus Game*, *Secret "X,"* and *Shell Game*, the contestant wins markers by guessing the prices of small items. The more markers he wins, the better the chance he wins the game prize corresponding to one of those markers. *Bonus Game* and *Shell Game*, which was initially designed to replace *Bonus Game*, are played infrequently and are less dynamic. *Cliffhanger* and *Secret "X"* are staples of *The Price Is Right* (*TPIR*) and offer more dramatic flair.

Game	Played	Wins	Losses	Win Rate	Average Prize	Median Prize	Avg Win	Median Win	Min Win	Max Win	Median Exp Win	Earnings Rank
Bonus Game	16	13	3	81%	$10,175	$ 9,819	$10,070	$ 9,458	$ 6,101	$18,264	$ 7,717	15
Cliffhanger	31	26	5	84%	$12,658	$ 9,305	$13,006	$ 9,996	$ 5,197	$72,044	$ 8,384	12
Secret X	30	16	14	53%	$10,501	$ 8,443	$11,563	$11,267	$ 6,930	$24,463	$ 6,034	29
Shell Game	10	7	3	70%	$ 8,940	$ 8,502	$11,577	$10,422	$ 5,758	$21,417	$ 7,338	18
Total/Wtd Avg	87	62	25	71%	$11,030	$ 9,010	$11,857	$10,259			$ 7,331	

Figure 17.1 Statistical Summary—Small Item Pricing Games

- *Bonus Game*: A board has four windows, each of which is associated with one item and one of which conceals the word "bonus." The contestant is shown four items, one at a time, each with a wrong price. He chooses whether the price of the item is higher or lower than the price shown. A correct choice means the corresponding window on the board is kept open. An incorrect choice means the window is closed. The "bonus" is then revealed. If adjacent to an open window, the contestant wins the game prize (i.e., the bonus prize). If adjacent to a closed window, the contestant loses. Players who correctly price all four items automatically win the bonus prize.

179

- *Cliffhanger:* A mountain climber climbs up an incline, which has unit markers one through twenty-five. The contestant is shown three items and guesses the price of each item, one at a time. The dollar difference between his guess and the actual price is the number of steps the mountain climber takes. The contestant wins the game prize if his guesses for three items are within $25 of the actual prices, so that the mountain climber does not take more than twenty-five steps and go over the cliff.
- *Secret "X":* The centerpiece is a large tic-tac-toe board. The middle column of the board contains a secret X hidden in one of its three boxes. To win the game prize, the contestant must get three X's in a row horizontally or diagonally. He starts with one free X to place on a left or right square and can win up to two more X's by guessing the price of sub-$100 items from two price options. He places the X's on the board, and the middle column is flipped over to see if he has a string of three X's in a row.
- *Shell Game:* There are four shells, one of which has a Ping-Pong ball underneath. The contestant can win up to four markers by guessing whether the price for each of four items is higher or lower than the price displayed for that item. He then places the markers in front of shells of his choosing. If a marker is in front of the shell with the Ping-Pong ball underneath, the contestant wins the game prize. If the contestant wins all four markers, he automatically wins. He also gets the chance to guess which shell has the Ping-Pong ball. If correct, he also wins the cash value of the game's prize.

BONUS GAME AND SHELL GAME

Bonus Game had an 81% win rate (13 of 16) and $7,700 median earnings, ranking it fifteenth among pricing games and among the best for non-car games. *Shell Game's* win rate of 70% (7 in 10) and $7,350 median earnings fell below that of *Bonus Game,* mainly due to chance. Contestants actually had better success pricing the small items in *Shell Game.*

What made the win rate so high in *Bonus Game* is the 67% success rate in winning the small items and keeping the windows open (i.e., in play for the bonus). In sixteen iterations of the game, the contestant kept all four windows open three times, three windows open eight times, two windows open three times, one window open one time, and zero windows open one time. Contestant guesses were balanced between guessing lower (34 times) and guessing higher (30 times).

Good luck also contributed to the 81% win rate. The player won the bonus two of the three times he won two items and the one time he won one item. Given the distribution of the number of items won, I would have expected an overall 67% win rate, as follows:

$$\left(100\% \times \frac{3}{16}\right) + \left(75\% \times \frac{8}{16}\right) + \left(50\% \times \frac{3}{16}\right) + \left(25\% \times \frac{1}{16}\right) + \left(0\% \times \frac{1}{16}\right)$$

$$= \frac{3}{16} + \frac{6}{16} + \frac{3}{32} + \frac{1}{64} = \frac{43}{64} = 67\%$$

Shell Game's $8,500 median prize value was below *Bonus Game*'s $9,800, but with the potential upside of doubling one's prize value in cash if winning all four markers and guessing which shell has the Ping-Pong ball underneath. Contestants earned all four markers four out of ten times and twice guessed which shell had the Ping-Pong ball, doubling their winnings. They earned three markers three times, winning once, and two markers three times, winning twice. Never did a contestant earn less than two markers. Given the distribution of markers earned, the 70% win rate fell a bit short of the expected win rate of 80%:

$$[(4 \times 100\%) + (4 \times 75\%) + (2 \times 50\%)]/10 = 80\%.$$

In *Shell Game*, contestants were 80% successful in guessing whether the price was higher or lower than the incorrect displayed price, even better than in *Bonus Game*. The first two markers were earned 100% of the time, the third marker 70% of the time, and the fourth marker 50% of the time. This divergence was partially due to chance but also reflected the third and fourth items having a price closer (in either direction) to the price displayed.* Similar to *Bonus Game*, contestants chose lower prices and higher prices for the small items with equal frequency, twenty-two times and eighteen times respectively.

CLIFFHANGER

A favorite among contestants and fans, *Cliffhanger* was first played in 1976. The median prize value was only $9,300, but the 84% win rate drove $8,400 median earnings, twelfth among pricing games.

Heuristics provides a virtually foolproof way of winning *Cliffhanger*. The first, second, and third items tend to be priced very closely to $25, $35, and $45, respectively. Guessing these values would have resulted in victory in

---

*The difference between the actual and displayed price was 47% for the first two small items, but only 34% for the last two small items.

twenty-nine out of twenty-nine iterations of *Cliffhanger*, and likely the other two instances in which the contestant lost before pricing the third item. In fact, the largest delta using this heuristic would have been $15, or fifteen steps for the mountain climber, with twenty-two out of the twenty-nine episodes leading to ten or fewer total steps. Given the foolproof and simple nature of this strategy, I see no reason to deviate from it.

The $25/$35/$45 heuristic is effective because the range of values for each of the three items is narrow. The first item ranged from $18 to $29; the second item ranged from $30 to $39; and the third item ranged from $40 to $49.

Cliffhanger, when lost, was typically due to one flawed guess. On November 7, 2018, a contestant lost on the third item after guessing $25 for a $49 water filter. On January 6, 2019, a contestant lost on the second item after guessing $75 for a $30 moisturizing system. And on October 1, 2019, the contestant found himself in a hole after guessing $44 for a $23 travel pillow on the first prize, sending the mountain climber up twenty-one steps. The remaining losses fit the same pattern: an extreme guess, sending the mountain climber up and over.

Although *Cliffhanger* is usually played for a sub-$10,000 prize, four times it was played for a car, including a $72,000 Jaguar F convertible during Dream Car Week on May 27, 2019. Three of the four car prizes were won on *Cliffhanger*, including the Jaguar, as the setup was no more difficult than usual.

Key Tip

Guess $25 for the first item, $35 for the second item, and $45 for the third item.

SECRET "X"

Secret "X" is *The Price Is Right*'s version of tic-tac-toe. There is a little skill, a dose of randomness, and ample suspense when the middle column is flipped over to reveal where the "secret *X*" is and whether the contestant has won.

If the contestant wins one *X*, he has two *X*s and a one in three chance of winning, since he can form only one row through one of the center boxes. If the contestant wins two *X*s, he should place all three *X*s in corner spots. He then has a two in three chance of winning by forming either a horizontal or a diagonal row through the middle with the three *X*s. Since he places the first *X* before winning the second *X* and the second *X* before winning the third, it follows that he should always pick the corners.

This strategy was borne out by the results of the game, as there were no mistakes in placement. The 53% observed win rate is consistent with the contestant winning one additional X in thirteen out of thirty games (43%) and two additional X's in seventeen out of thirty games (57%), which would predict a 52% win rate: $\left(\dfrac{1}{3} \times 43\% \right) + \left(\dfrac{2}{3} \times 57\% \right) = 52\%$.

Contestants won forty-seven out of sixty additional X's across thirty pricing games (78%) by selecting the correct price for the small item from two displayed prices. Driving the 78% success rate was the wide disparity between the high displayed price and low displayed price. The first small item had an average high price of $72, an 85% premium to the average low price of $39. The contestant won the first item and an additional X an impressive 87% of the time (26 out of 30). The two price options for the second item were closer together, with a 62% premium between the average high price ($115) and the average low price ($71). Consequently, the second X was won only 70% of the time (21 out of 30).

The contestant guessed the low price for the small items forty times (67%) and the high price twenty times (33%). The bias to guess low seemed more pronounced for unfamiliar items, similar to Contestant's Row bidding, yet the correct pick was just as likely to be the high price as the low price.

## 18

Car Pricing Games, Part 1

There were twenty-eight car games on *The Price Is Right* (*TPIR*) during seasons 47 and 48. Almost without fail, there are two car games per show, one in the first half of the show and one in the second half. In addition, a car is always offered in one showcase and sometimes in both. Sixteen car games are car pricing games, in which the contestant guesses the price of the car in some fashion and there are no other material prizes at stake. All car games, including car pricing games, are played most frequently in pricing games three, five, and six, with roughly 20% of all car games falling in each segment, and fewer in pricing games one, two, and four.

Given the median $20,000 prize value and win rates in the 30 to 50% range, car pricing games lead to high median earnings, averaging $7,500 across games, as shown in Figure 18.1. Five car pricing games ranked in the top ten

Game	Played	Wins	Losses	Win Rate	Average Prize	Median Prize	Avg Win	Median Win	Min Win	Max Win	Median Earnings	Earnings Rank
Identifying Car Digits												
Cover Up	39	16	23	41%	$20,995	$20,814	$21,239	$20,430	$16,425	$31,428	$ 8,382	13
Gridlock	24	11	13	46%	$20,369	$20,500	$19,975	$20,413	$15,863	$21,938	$ 9,356	10
Line 'em Up	23	12	11	52%	$19,461	$19,561	$19,465	$19,500	$17,561	$20,995	$10,174	8
Money Game	46	25	21	54%	$20,286	$20,616	$19,820	$19,891	$16,352	$23,211	$10,905	6
One Away	32	8	24	25%	$23,211	$20,903	$25,481	$21,418	$18,257	$35,271	$ 5,355	37
Pathfinder	24	5	19	21%	$20,632	$20,668	$19,713	$19,738	$17,593	$23,615	$ 4,242	53
Pocket Change	19	7	12	37%	$22,754	$20,613	$19,805	$19,823	$16,837	$21,654	$ 7,303	19
Stack the Deck	18	3	15	17%	$21,405	$20,668	$23,103	$24,183	$19,478	$25,649	$ 4,031	57
Ten Chances	10	4	6	40%	$12,455	$19,645	$20,000	$20,070	$19,500	$20,360	$ 8,518	11
Car Digits Versus Other Prizes												
Any Number	42	12	30	29%	$22,432	$21,446	$21,048	$21,559	$16,257	$24,841	$ 6,453	26
Switcheroo	19	4	15	21%	$19,791	$20,120	$20,029	$20,718	$17,315	$21,365	$ 4,362	48
Car Digits: Over/Under												
Dice Game	33	16	17	48%	$19,492	$21,225	$19,991	$21,308	$15,345	$24,345	$10,331	7
Lucky Seven	55	17	38	31%	$20,764	$21,492	$21,045	$21,593	$17,298	$25,645	$ 6,674	23
Getting Close in Price												
Card Game	31	11	20	35%	$22,866	$22,885	$22,052	$22,070	$21,111	$23,496	$ 7,831	14
Five Price Tags	19	10	9	53%	$25,862	$20,516	$26,083	$20,849	$17,388	$79,791	$11,014	4
That's Too Much	49	11	38	22%	$24,910	$22,852	$25,605	$23,300	$19,291	$51,700	$ 5,231	38
Total/Wtd Avg	483	172	311	36%	$21,517	$21,142	$21,363	$20,939			$ 7,509	

Figure 18.1 Statistical Summary—Car Pricing Games

for median earnings: *Dice Game, Five Price Tags, Gridlock, Line 'Em Up,* and *Money Game.* Another four ranked in the top twenty: *Card Game, Cover Up, Pocket Change,* and *Ten Chances.*

Which car pricing game a contestant plays has a significant bearing on his chances of winning a car. Contestants rarely won *One Away, Pathfinder, Stack the Deck, Switcheroo,* and *That's Too Much,* and for good reason. The odds are stacked against the player. On the opposite end of the spectrum, *Ten Chances* has about 90% odds of success if played correctly, whereas *Cover Up, Dice Game,* and *Money Game,* if played correctly, have better than 50% odds of success.

Most of *TPIR's* car pricing games had median prize values in the $19,000 to $22,000 range. Exceptions on the high end were *Card Game* and *That's Too Much,* each with a $22,900 median car value that added a degree of difficulty. In each of seasons 47 and 48, about two-thirds of car pricing games offered cars between $18,000 and $23,000. Contestants won as frequently with cars priced more than $21,000—36% of the time—as they did with cars less than $21,000. Of the 483 cars offered in car pricing games, 29 had a retail price of more than $25,000.

Here are the rules for the nine of the car pricing games that focus on identifying car digits. The other seven car pricing games are covered in Chapter 19.

IDENTIFYING CAR DIGITS

The contestant picks individual or clusters of digits in the car price.

- *Cover Up*: A contestant must choose each digit in the price of the car from ascending columns of number choices, with two options for the first digit, three for the second digit, four for the third digit, five for the fourth digit, and six for the fifth digit. If, after the contestant's first choice, at least one digit is correct and lights up, the game continues with the contestant choosing other options for the incorrect digits. If the contestant gets at least one incremental digit correct in his revised guess, the game continues. The game ends with the contestant winning the car or not having any incremental numbers correct in his next guess.
- *Gridlock*: The contestant is shown a series of mini cars and is given the first digit in the price with the gray car. Next, he picks a pair of numbers from one of three blue cars to guess the second and third digits in the car. He then picks a second pair of digits from one of three orange cars to guess the fourth and fifth digits in the car. The contestant gets one do-over and loses on a second incorrect guess.
- *Line 'Em Up*: The contestant is shown a vertical game board with the first and last digits in the price of a car. Between these digits are the prices of three smaller prizes stacked horizontally, each of which contains one

of the other three digits in the price of the car. The contestant slides these prices horizontally so that the digits in the price of the car line up vertically in a yellow frame. The first and third prizes have three-digit prices, and the second prize has a two-digit price. If the contestant correctly lines up all of the car's middle digits from the three prizes, he wins the car and the small prizes. Otherwise, the contestant is told how many but not which of the digits are correctly placed. He then has one more attempt to correctly line up the digits to win the car and smaller prizes.

- *Money Game*: The contestant is shown a three-by-three board of nine two-digit numbers and the middle digit in the five-digit car price. One of the nine two-digit numbers is the first two digits in the price of the car (front half of the car) and one is the last two digits in the price of the car (back half of the car). The contestant wins the amount of money equal to the two-digit number for each incorrect guess. In order to win, the contestant must choose the front and back half digit pairings before choosing four money tags, meaning he can only err three times.

- *One Away*: The contestant is shown a five-digit number, with each digit one away from the number in the car price. He decides for each digit whether the corresponding number in the price of the car is one number higher (colored in blue) or one number lower (colored in red). He then asks "Mighty Sound Effects Lady" backstage if he has at least one number correct, two numbers correct, and so forth, with the horn sounding until he fails to get an additional correct number. Upon knowing how many but not which numbers he has correct, he has one opportunity to adjust (i.e., higher to lower or lower to higher) numbers in the price of the car. If his second guess is correct, he wins the car.

- *Pathfinder*: The contestant stands in the middle of a five-by-five board of numbers. The middle number represents the first digit in the price of the car. He then moves to one of the four adjacent numbers that he believes is the second digit, one of the three numbers that he believes is the third digit, one of the two or three numbers that he believes is the fourth digit, and one of two numbers that he believes is the fifth digit, never repeating squares. An incorrect guess requires that he pick the price of a small item (i.e., sub-$200) from two possible price choices. He can win up to three do-overs with correct picks. The contestant either guesses the price of the car and wins or runs out of do-overs and loses.

- *Pocket Change*: The contestant is given $0.25, the "selling" price of a car. The contestant is also shown a game board, which has spaces for the five digits in the price of the car and six numbers lit up in circles at the top of the game board. The first digit in the price is revealed and removed from the six lit circles above, leaving four of five numbers to place. The contestant then selects one of the remaining numbers as the second digit in the car. If he is correct, the number is revealed in the price and removed from the remaining five number choices. The contestant also receives his choice of one of twenty pouches attached to the game board, which contain various amounts of "pocket change." If his guess is incorrect, the selling price of the car increases by $0.25 and he guesses again with the same consequences. Play continues this way for each of the remaining digits in the price of the car. Once the car is priced, the contents of the selected pouches are revealed to see if the contestant has earned enough pocket change to buy the car at the "selling price."

- *Stack the Deck*: The contestant is shown seven different digits (in the style of playing cards), five of which are digits in the price of a car. The contestant is given an opportunity to "stack the deck" in his favor and receive up to three digits in the price of the car. The contestant is shown two grocery items with a displayed price and must select which item corresponds to the price. The contestant has three such opportunities. For each correct answer, the contestant chooses one of the digits in the car's price to be revealed. The contestant must then try to guess the remaining numbers correctly using the not-yet-placed cards. The car price is then revealed to see if the contestant wins.

- *Ten Chances*: One of the slower games, the contestant is shown a two-digit prize, a three-digit prize, and a car prize. The contestant has a maximum of ten chances to guess the price of three prizes: first the sub-$100 prize using two of three digits, second the sub-$1,000 prize using three of four digits, and finally the car using all five digits provided. He wins if he guesses the price of the car before using all ten chances. Otherwise, he wins the smaller prizes he has guessed correctly.

COVER UP: EXTENDING THE GAME

A course in probability would do wonders for a contestant in *Cover Up*. Although performance was good with a 41% win rate (16 out of 39) and $8,400 median earnings, performance could have been much better. The fundamental tension in *Cover Up* is between getting new digits in the car price correct and getting so many new digits in the car price correct that one limits his future chances. For example, if the contestant is fortunate enough to guess the first four digits correctly on his first turn, he only has one turn with which to guess the last digit correctly from the remaining five choices. Not good odds at all. Figure 18.2

Games	1st Try	2nd Try	3rd Try	4th Try	5th Try	Total
Win	0	5	6	3	2	16
Loss	1	12	10	0	0	23
Total	1	17	16	3	2	39

Figure 18.2 *Cover Up* **Wins and Losses by Try**

shows the number of wins and losses by turn in *Cover Up*, which is illustrated in Figure 18.3. If the contestant reached his third try, he won eleven out of twenty-one times (sum of third, fourth, and fifth turns) and was five for five if he got to a fourth try.

What no contestant did during seasons 47 and 48, which is probabilistically superior, is to purposefully miss the first digit in the car so as to get more chances at digits two through five. Since the first digit is known with certainty, it can be missed with certainty. Only one contestant missed the first digit, and the miss did not appear intentional. And only three times did the contestant get only the

Figure 18.3 *Cover Up.* **Illustration courtesy of Omar Faruk**

first digit right on his first try. The other thirty-five times he got at least one other digit correct on his first try. Had he purposefully missed the first digit, the game would have continued thirty-five out of thirty-nine times (90%), ensuring at least three tries for the car (with the first digit always correct on the second try).

Figure 18.4 shows success in getting car digits correct across one's various tries during seasons 47 and 48, as both a fraction and a percentage. Observe that on the first try, the second, third, fourth, and fifth digits were picked 77%, 41%, 21%, and 15% of the time respectively. Note that it would have been possible to have had a third try for the second digit, a fourth try for the third digit, and a fifth try for the fourth digit.[*] It just never transpired.

Car Digit Try	1st	2nd	3rd	4th	5th	1st	2nd	3rd	4th	5th
1	38/39	30/39	16/39	8/39	6/39	97%	77%	41%	21%	15%
2		8/8	9/22	9/30	9/32		100%	41%	30%	28%
3			3/7	4/11	4/14			43%	36%	29%
4				2/2	1/3				100%	33%
5					2/2					100%

Figure 18.4 *Cover Up.* **Success by Try and Car Digit**

To demonstrate the superiority of missing the first digit, I use a probabilistic model. Assume for purposes of approximation that the first digit can be picked correctly 100% of the time, the second digit 80% of the time (8 in 10), the third digit 33% of the time (1 in 3), the fourth digit 25% of the time (1 in 4), and the fifth digit 20% of the time (1 in 5). For the second digit, 80% assumes an informed guess and compares to the 77% observed in seasons 47 and 48. For digits 3–5, the probabilities assume a player can eliminate one of the choices and make a random guess from the remaining selections. So for the fourth digit, the player is effectively choosing from four options after ruling out the fifth option. The chance of missing digits 2–5 on one's first try in this model is $\frac{1}{5} \times \frac{2}{3} \times \frac{3}{4} \times \frac{4}{5} = \frac{2}{25}$, or 8%. So intentionally missing the first digit would lead to a *modeled* loss after the first guess 8% of the time versus the 10% that would have transpired had all contestants intentionally missed the first digit in seasons 47 and 48.

Dozens of scenarios make the subsequent analysis rather complicated, and one needs to come up with representative probabilities for each round.

---

[*]Subsequent tries have a smaller number in the denominator on account of the contestant (a) having lost the game on a prior try, or (b) having gotten the digit in question correct on a prior try.

That is, in the second round, if the contestant has not correctly chosen digit 2, 3, 4, or 5 in the first round, what are the probabilities of getting each of the digits correct on the second try, with one possible selection now eliminated? To model this, I mapped out the full set of scenarios with the sample probabilities in Figure 18.5, treating the first digit as being missed with certainty or made with certainty. If the contestant purposefully selects the correct first digit, there are 193 possible paths in the game ending in one, two, three, four, or five tries, some winning paths and some losing paths, and each with its own probability. If the contestant intentionally selects the incorrect first digit, there are 178 possible paths (winning and losing), again each with its own probability.

Car Digit Try	1st Miss	1st Make	2nd	3rd	4th	5th
1	0%	100%	80%	33%	25%	20%
2			75%	50%	33%	25%
3			100%	60%	40%	30%
4				100%	60%	40%
5						60%

Figure 18.5 *Cover Up* **Probabilistic Modeling—Sample Probabilities**

With the probabilities noted in Figure 18.5, the contestant would have a 47% chance of winning using the strategy of purposefully missing the first digit, as seen in Figure 18.6. However, he would only have a 36% chance of winning when choosing the first digit correctly on the first try. Figure 18.6 shows, after each try, the player's probability of winning (W), losing (L), or continuing to play (C) another round.

Intentionally Miss First Number

Try	1	2	3	4	5
W	0%	12%	33%	45%	47%
L	8%	8%	43%	52%	53%
C	92%	80%	24%	3%	0%

Intentionally Make First Number

Try	1	2	3	4	5
W	1%	13%	29%	35%	36%
L	0%	33%	59%	64%	64%
C	99%	55%	13%	1%	0%

W = Win   L = Loss   C = Continue to Next Try

Figure 18.6 *Cover Up* **Expected Win Rates—Missing versus Making First Digit**

The major benefit of intentionally missing the first digit is to increase one's chances of extending the game to at least four rounds. If the contestant gets to a fifth round, he is either playing for a sure bet on the fourth number (because the other four choices have been eliminated) or at least a 50/50 chance on the fifth number (because four of the six choices have been eliminated). Importantly, purposefully missing the first digit—assuming one correct digit in the first round—ensures continuing play for a third round. Making the first digit, however, creates material risk of elimination in the second round, when the player might have to guess one of the more difficult digits correctly (i.e., digits 3–5) to stay in the game.

The benefit from intentionally missing the first digit should hold as long as the probability of success on the second digit is relatively high. The magnitude of the benefit increases the closer to certain one is on the second digit, with certainty on the second digit ensuring the contestant has at least three tries when *intentionally missing* the first digit. Conversely, near certainty on the second digit lowers the odds of success when *intentionally making* the first digit. The player will likely also get the second digit correct on his initial try, limiting his tries for digits 3–5, which have many choices and are essentially a random guess.

Extending the Game

The strategic imperative in *Cover Up* is to extend the game to multiple rounds. There are many examples in life in which success requires one to extend a game to multiple rounds. Job interviews are often three rounds. The goal in the first round is to impress one or more interviewers and progress to the second round. Likewise for the second round, with hiring then tied to a positive consensus after two or three rounds. It is rare to get hired after one round, and the benefit from doing exceptionally well with one interviewer means little if the other interviewers do not agree. In a negotiation, it is best to seek agreement on a few issues at a time. Reaching agreement on most of the issues up front can lead to an impasse when tackling the remaining, often thornier issues. Extending the negotiation, on the other hand, means multiple chances to resolve stickier issues. Talks in later rounds often get easier as the two parties gain comfort resolving their differences. Consider the dialogue in Congress in 2021 around the Infrastructure and Jobs Act. The sticking point was how tax revenue would be raised to pay for the bill. Agreeing first on the spending priorities and next on the amount of spending meant multiple tries on how to pay for such a bill.

Finally, consider heuristics for digit 5. I tested historically observed patterns through Monte Carlo simulations to evaluate the likelihood of the following heuristics occurring through chance. I ran thousands of simulations of up to thirty-nine trials to see how frequently or infrequently a number would appear. Key conclusions were as follows:

- For the fifth digit, avoid choosing 5 or 7, which were the correct choice one in thirty-three times and zero in twenty-three times, respectively. There is a less than 10% random chance that any number would show up as infrequently as 5 and 7 showed up as the fifth digit.
- For the fifth digit, prioritize digits 3 and 8, which were the correct choice nine in seventeen times (53%) and twelve in twenty-two times (55%), respectively. A number would have shown up as frequently as 3 did only 2% of the time by random chance and even less for 8.

Cover Up heuristics are beneficial regardless of whether the contestant purposefully misses or makes the first digit on his first try. But the heuristics listed here should further improve one's win rate above 50% (from 47%) if purposefully missing the first number and assuming the probabilities in Figure 18.5.

Key Tip

Intentionally miss the first digit on the first try to increase the number of likely tries in *Cover Up*. Note the heuristics, particularly the likelihood of 3 or 8 as fifth digit options.

GRIDLOCK

Gridlock is a game of limited strategy, but with a high observed win rate—46% (11 out of 24), leading to solid median earnings of $9,350 (tenth across all pricing games). It is the newest *TPIR* pricing game, introduced in 2017. There are three ways to win *Gridlock*:

1. Get both pairs of digits correct on the first try (6 wins).
2. Get the second and third digit pairing correct after one miss and the fourth and fifth digit pairing correct on one's first try (2 wins).
3. Get the second and third digit pairing correct on one's first try and the fourth and fifth digit pairing correct after one miss (3 wins).

Although this might seem like a difficult task, there is usually one option for the second and third digit pairing that is less likely. Random chance points to a 33% chance of winning *Gridlock*. Better than random odds on the second and third digit pairing—for example 50% odds of success on one's first try and 60% odds on one's second try—increases one's chances of victory to about 46%. During seasons 47 and 48, contestants got the second and third digit pairing correct 46% of the time on their first try (11 out of 24) and 54% of the time on their second try (7 out of 13).

The 46% modeled win rate matches the 46% observed win rate. The contestant won 25% of the time with no mistakes, 8% of the time with a mistake on his first pairing, and 13% of the time with a mistake on his second pairing.

Heuristically, the second and third digit pairing were only the high choice once versus eight times for the middle choice and fifteen times for the low choice. If the choices were truly random, a simulation of twenty-four trials would lead to a choice (high, medium, or low) coming up zero or once only one in every four hundred simulations. Interestingly, the high choice for the second and third digits was correct five out of fifteen times during season 49, indicating that the showrunners may have since altered course. So I do not highlight this as a key tip.

LINE 'EM UP

With a 52% win rate (12 out of 23), a $19,550 median prize value, and $10,150 median earnings, *Line 'Em Up* ranked eighth in median earnings and fourth among car pricing games, although mainly due to good luck. During seasons 47 and 48, players won four times on their first attempt and eight times on their second attempt.

The game may seem daunting at first glance. Three numbers for the second digit, two possible numbers for the third digit, and three possible numbers for the fourth digit mean $3 \times 2 \times 3 = 18$ possible combinations. Though the contestant gets two chances and can usually eliminate one wrong choice for the second number in the car, that still leaves twelve possible combinations. Yet contestants won more than half the time.

Figure 18.7 captures the episode that aired June 5, 2019. The contestant was unsuccessful guessing the price of the $16,503 Ford Fiesta. The choices for the second digit were all one number apart, making for a very hard selection, although $15,000 may have seemed a bit too low for a car on *TPIR*.

Strategically, there is not much at play in *Line 'Em Up*. But a sensible choice on the second digit and heuristics on the fourth digit can be quite helpful. If the player can rule out a second digit and fourth digit option with near certainty, possible combinations would be narrowed to eight (two for the second number, two for the third, and two for the fourth, multiplied together). This is a realistic model for the game.

The second digit varied in terms of its degree of difficulty. Thirteen out of twenty-three times the contestant had to choose from a continuous sequence of numbers. Three times the possible digits were two apart (e.g., 5, 7, 9), and seven times two of the digits were one apart and the other two were two apart (e.g., 6, 8, 9). In only four out of twenty-three iterations of the game

Figure 18.7 *Line 'Em Up. Illustration courtesy of Omar Faruk*

was the second digit the lowest of the three options. Seven times the second digit was the highest choice, and twelve times it was the middle number. Interestingly, the fourth digit in the car was 5 only once in fourteen times when a 5 was in play, and never 9 despite ten choices with 9 as an option.

What are the odds of winning *Line 'Em Up*? I evaluate three scenarios:

1. The contestant guesses at random.
2. The contestant is able to reject one of the choices for the second digit (assume it's the low number).
3. The contestant is able to reject one of the choices for the second digit and one of the choices for the fourth digit.

One's probability of winning in *Line 'Em Up* is much better than randomly picking from the different options. Why? Because being given the number of correct digits after one's first guess reduces the number of possible guesses on one's second attempt. First consider the random guess scenario, shown in Figure 18.8. The contestant has eight possible outcomes on his first

Random Guess

	2nd Digit	3rd Digit	4th Digit	# Right	Odds: 1st Outcome	Odds: 2nd Guess Win	Prob Win	Prob Win %
All correct	1	1	1	3	1/18	1	1/18	5.6%
2 digits correct	1	1	0	2	1/9	1/5	1/45	2.2%
	1	0	1	2	1/18	1/5	1/90	1.1%
	0	1	1	2	1/9	1/5	1/45	2.2%
1 digit correct	1	0	0	1	1/9	1/8	1/72	1.4%
	0	1	0	1	2/9	1/8	1/36	2.8%
	0	0	1	1	1/9	1/8	1/72	1.4%
0 correct	0	0	0	0	2/9	1/4	1/18	5.6%
					100%		2/9	22.2%

Eliminate One 2nd Number Choice

	2nd Digit	3rd Digit	4th Digit	# Right	Odds: 1st Outcome	Odds: 2nd Guess Win	Prob Win	Prob Win %
All correct	1	1	1	3	1/12	1	1/12	8.3%
2 digits correct	1	1	0	2	1/6	1/4	1/24	4.2%
	1	0	1	2	1/12	1/4	1/48	2.1%
	0	1	1	2	1/12	1/4	1/48	2.1%
1 digit correct	1	0	0	1	1/6	1/5	1/30	3.3%
	0	1	0	1	1/6	1/5	1/30	3.3%
	0	0	1	1	1/12	1/5	1/60	1.7%
0 correct	0	0	0	0	1/6	1/2	1/12	8.3%
					100%		1/3	33.3%

Eliminate One 2nd Number, One 4th Number Choice

	2nd Digit	3rd Digit	4th Digit	# Right	Odds: 1st Outcome	Odds: 2nd Guess Win	Prob Win	Prob Win %
All correct	1	1	1	3	1/8	1	1/8	12.5%
2 digits correct	1	1	0	2	1/8	1/3	1/24	4.2%
	1	0	1	2	1/8	1/3	1/24	4.2%
	0	1	1	2	1/8	1/3	1/24	4.2%
1 digit correct	1	0	0	1	1/8	1/3	1/24	4.2%
	0	1	0	1	1/8	1/3	1/24	4.2%
	0	0	1	1	1/8	1/3	1/24	4.2%
0 correct	0	0	0	0	1/8	1	1/8	12.5%
					100%		1/2	50.0%

Figure 18.8 *Line 'Em Up* Probabilities

guess, one of which involves getting all numbers correct, one of which involves missing all numbers, three of which involve getting one number correct, and three of which involve getting two numbers correct. Each of those eight scenarios has a certain number of combinations out of the eighteen possible combinations ($3 \times 2 \times 3$) that map to that outcome, which drives the probability of that outcome (i.e., the number of combinations out of 18).

Now consider the second attempt. If the contestant gets all numbers correct on his first try, he wins. If he gets two numbers correct (and he does not know which two), then he must change one of the numbers. This involves two possible alternative choices for the second number, one alternative for the third number, and two for the fourth digit, for a one in five, or 20% chance of getting the car price correct.

If he gets one number correct (and he does not know which one), then he has to change two of the numbers, which involves eight possible choices: if the second number is correct, there is one alternative for the third and two for the fourth (1×2); if the third number is correct, there are two alternatives for the second and two for the fourth (2×2); if the fourth number is correct, there are two alternatives for the second and one for the third (1×2)—for a total of eight possible choices, and a one in eight chance of winning. If all numbers are wrong, the third number is known (the other choice), but there are two possible choices for the second number and two for the fourth number, for a total of four possible choices, and a one in four chance of winning.

In the second scenario in Figure 18.8, I assume one choice for the second number can be eliminated. There are still eight outcomes, similar to random chance, but those outcomes encompass only twelve first-choice combinations ($2 \times 2 \times 3$). There are five second-choice combinations if one digit is correct and four second choices if two digits are correct. If the contestant can eliminate one second-number option and one fourth-number option (Figure 18.8, third scenario), he is left with only eight first-choice combinations and even fewer conditional second-choice combinations.

Eliminating one of the second digit choices with certainty is the most representative scenario for a contestant on the show, as he is likely unaware of 5 and 9 being unlikely for the fourth digit. It is also representative of the illustrated game, with 5 too low for the second digit. The contestant may have felt the same way, guessing $17,583 for the second digit in the Ford Fiesta. At that point, he had one digit correct, with five possible choices.

1. Change the second digit to a 6 and the third digit to a 2. Keep the fourth digit as an 8: $16,283.
2. Change the second digit to a 6 and the fourth digit to a 7. Keep the third digit as a 5: $16,573.

3. Change the second digit to a 6 and the fourth digit to a 0. Keep the third digit as a 5: $16,503.
4. Keep the second digit as a 7. Change the third digit to a 2 and the fourth digit to a 7: $17,273.
5. Keep the second digit as a 7. Change the third digit to a 2 and the fourth digit to a 0: $17,203.

The contestant went with $16,573 and lost, as the right answer was $16,503.

The 52% win rate (12 out of 23) during seasons 47 and 48 was much better than the modeled 33% win rate, and more akin to the contestant eliminating both a second digit choice and a fourth digit choice. But this success was likely due to chance along with the contestant eliminating one number choice. Not only did the contestant get the car price correct four times on his first attempt in twenty-three tries, but he also won four of five times on his second attempt after getting all digits wrong on his first try. With one second-digit choice eliminated, these outcomes would have been expected to occur one-twelfth of the time and half the time respectively.

Key Tip

Try to eliminate one option for the second and fourth digits. The low number choice is unlikely to be the second digit and 5 or 9 are unlikely to be the fourth digits.

MONEY GAME

Money Game is a classic *Price Is Right* car game. Contestants won twenty-five times in forty-six games for a 54% win rate, which when coupled with a median $20,000 car, led to $10,900 median earnings, ranking second among car pricing games.

Odds are quite good in *Money Game* if the contestant does not waste picks. Almost every *Money Game* has three realistic values for the first two digits in the car price, leaving six numbers as possible values for the last two digits in the car. (Recall the third digit is revealed.)

Consider the episode that aired June 24, 2019. The three possible numbers for the first two digits for a Mazda 3 Sport were 18, 19, and 20. The other six number choices were possibilities for the last two digits, including 85, 73, 47, 96, 54, and 68. The contestant correctly guessed the first two digits on his first try and had four tries for the last two digits. He got the last two digits, 68, correct on the third try after missing with 85 and 73, winning the $20,768 car.

Playing *Money Game* well is straightforward and so are the odds. First, guess the first two digits from the three relevant choices. Once you guess the front of the car, guess the last two digits, avoiding the remaining possibilities for the first two digits.

Now consider the probabilities of winning:

- At least one-third of the time, the player will get the first two digits correct on his first try. He will then have a two-thirds chance of winning the car (4 selections from 6 choices for the last two digits).
- One-third of the time or less, the player will get the first two digits correct on his second try. He will then have a one in two chance of winning the car (3 selections from 6 choices).
- One-third of the time or less, the player will get the first two digits correct on his third try. He will then have a one in three chance of winning the car (2 selections from 6 possible choices).

The probability of a win: $P(\text{win}) > \left(\frac{1}{3} \times \frac{2}{3}\right) + \left(\frac{1}{3} \times \frac{1}{2}\right) + \left(\frac{1}{3} \times \frac{1}{3}\right) = \frac{2}{9} + \frac{1}{6} + \frac{1}{9} = \frac{1}{2} = 50\%$.

If the contestant gets the first two digits correct on the first try more than one-third of the time, the probability of winning increases. For example, if the odds on the first two digits across tries one through three are 45%/30%/25%, then the odds of winning increase to $P'(\text{win}) = \left(\frac{9}{20} \times \frac{2}{3}\right) + \left(\frac{3}{10} \times \frac{1}{2}\right) + \left(\frac{5}{20} \times \frac{1}{3}\right) = \frac{3}{10} + \frac{3}{20} + \frac{1}{12} = 53\%$. Contestants correctly guessed the first two digits on their first try twenty-one out of forty-five times.[†] Eleven of the other twenty-four contestants were correct on their second try, with the remaining thirteen correct on their third try.

It is not easy to err in *Money Game*. Still, after getting the first two digits correct, five players chose numbers more appropriate for the front of the car, effectively wasting a guess on the back half of the car. All five lost.

Heuristics can also improve a contestant's chances on the last two numbers. The number 47 was an option in every *Money Game* during season 47, and the number 48 in every game during season 48. Each number corresponded to the season at hand. Not once was that number in the car's price. What an amusing heuristic! Contestants seemed completely oblivious, picking 47 or 48 eighteen times, which represented close to one-sixth of their tries when targeting the back

---

[†]Excludes one game in which the contestant did not initially focus on the first two digits.

half of the car.‡ Had contestants avoided numbers 47 and 48, they would have had only five choices for the last two numbers, and their odds per guess would improve to one in five from one in six. Odds per guess on the first two numbers would remain one in three. The overall chance of victory would therefore improve to $P(\text{win with heuristics}) > \left(\frac{1}{3} \times \frac{4}{5}\right) + \left(\frac{1}{3} \times \frac{3}{5}\right) + \left(\frac{1}{3} \times \frac{2}{5}\right) = \frac{4}{15} + \frac{1}{5} + \frac{2}{15}$ $= \frac{9}{15} = 60\%$. With better than random chance for the first two digits (45%/30%/25%), odds further improve to $P'(\text{win with heuristics}) = \left(\frac{9}{20} \times \frac{4}{5}\right)$ $+ \left(\frac{3}{10} \times \frac{3}{5}\right) + \left(\frac{5}{20} \times \frac{2}{5}\right) = \frac{9}{25} + \frac{9}{50} + \frac{1}{10} = \frac{32}{50} = 64\%$.

Money Game can be one of the winningest car games if players take advantage of observed heuristics.

Key Tip

If past is prologue, avoid the number corresponding to the season number of *TPIR*.

ONE AWAY

One Away is embellished by the contestant asking, "Mighty Sound Effects Lady, do I have at least one number correct?" (or at least two, at least three, at least four, or all five numbers correct) after initially guessing the car's price. What *One Away* achieves in drama is lost in contestant success. The 25% win rate (8 out of 32) and $20,900 median prize value led to $5,350 median earnings, thirty-seventh across pricing games and bottom quartile among car pricing games. Only *Pathfinder*, *Stack the Deck*, and *Switcheroo* ranked lower.

The setup in *One Away*, illustrated in Figure 18.9, is most analogous to *Line 'Em Up*. There are probabilities for one's first pick and then conditional probabilities for the second pick based on how many digits the contestant gets right in his first guess. Contestants never got the first digit wrong in their first

---

‡Looking beyond seasons 47 and 48, www.tpirstats.com indicates that the numbers 45, 46, and 49 were options in every *Money Game* in seasons 45, 46, and 49, respectively; 45 was the right choice once in twenty-seven games, 46 once in twenty-eight games, and 49 never in seventeen games.

Figure 18.9 One Away. Illustration courtesy of Omar Faruk

guess, which is not surprising. On the other hand, digits 2–5 functioned like random guesses, with 65 out of 128 correct picks, as follows:

	2nd Digit	3rd Digit	4th Digit	5th Digit
Correct	16 out of 32	19 out of 32	13 out of 32	17 out of 32

Based on this, *One Away* can be thought of as analogous to a game in which one flips four coins with the goal of getting four heads (corresponding to numbers 2–5 in the car's price) and is then told how many heads were flipped and given the opportunity to change the number of tail flips without knowing which coins were flipped tails. Here, the chance of a successful first guess is $1/2 \times 1/2 \times 1/2 \times 1/2 = 1/16$. Assuming an unsuccessful first attempt, the odds are not random on the second try, because the contestant only flips over the number of coins corresponding to the number of non-heads. So we can consider a set of first guess outcomes and conditional win probabilities, similar to *Line 'Em Up*, which then add up to the total win probability.

The analysis shown in Figure 18.10 is actually far simpler than in *Line 'Em Up*, as the probabilities are simpler. There is a one in sixteen chance of each outcome in one's first guess, and a one in four chance of correctly changing the incorrect number if one gets all but one of digits 2–5 correct initially (changing either digit 2, 3, 4, or 5). Ditto for only getting one of digits 2–5 correct. Getting two of digits 2–5 correct means six possible choices for digits to change (digits 2 and 3, 2 and 4, 2 and 5, 3 and 4, 3 and 5, or 4 and 5).

Random Guess

2nd Digit	3rd Digit	4th Digit	5th Digit	# Right	Prob 1st Outcome	Prob Win 2nd Try	Prob Win	Prob Win %
1	1	1	1	4	1/16	1	1/16	6.3%
1	1	1	0	3	1/16	1/4	1/64	1.6%
1	1	0	1	3	1/16	1/4	1/64	1.6%
1	0	1	1	3	1/16	1/4	1/64	1.6%
0	1	1	1	3	1/16	1/4	1/64	1.6%
1	1	0	0	2	1/16	1/6	1/96	1.0%
1	0	1	0	2	1/16	1/6	1/96	1.0%
1	0	0	1	2	1/16	1/6	1/96	1.0%
0	1	1	0	2	1/16	1/6	1/96	1.0%
0	1	0	1	2	1/16	1/6	1/96	1.0%
0	0	1	1	2	1/16	1/6	1/96	1.0%
1	0	0	0	1	1/16	1/4	1/64	1.6%
0	1	0	0	1	1/16	1/4	1/64	1.6%
0	0	1	0	1	1/16	1/4	1/64	1.6%
0	0	0	1	1	1/16	1/4	1/64	1.6%
0	0	0	0	0	1/16	1	1/16	6.3%
1 = Correct			*0 = Incorrect*				5/16	31.3%

Figure 18.10 *One Away* Probabilities

My model predicts a 31% chance of success, not far off from the 25% win rate observed during seasons 47 and 48. We can also compare the number of digits correct in one's first guess to the model's prediction.

Digits 2–5 Correct	0	1	2	3	4
# Games, Seasons 47 and 48	2	6	13	10	1
# Games, Modeled	2	8	12	8	2

Are there heuristics that can make *One Away* easier? Yes. There is one essential heuristic. *Never choose the same number for more than one of digits 2–5.* Four of thirty-two first guesses and seven of thirty-one second guesses featured a duplicated number, yet the car price never had a duplicated number in digits 2–5. Eleven out of sixty-three guesses were effectively wasted. In *One Away*, the average number of possibilities with duplicated numbers in digits 2–5 was 5.3 out of 16 possible combinations. Some games had as many as eleven or twelve choices with duplicated numbers, allowing the player to gain a meaningful advantage, while four games had no duplicated possibilities, offering no advantage.

Consider the number string 28643. The nonrepeating number heuristic means the contestant should not move the second digit one down (D) to 7 and the third digit one up to 7 (U) or the third digit one down (D) to 5 and the fourth digit one up (U) to 5 when guessing the price of the car. Normally there would be sixteen possible combinations, but the nonrepeating number strategy means there will never be a first down (D) followed by a second up (U) or a second down (D) followed by a third up (U). I can therefore eliminate $2 \times 2 = 4$ choices for each of two overlapping number combinations, leaving eight suitable choices. With eight choices, there are 57.3% odds of success on one's second try, assuming each of the eight suitable choices is equally likely. With a one in eight chance of success on the first try, an informed contestant would win *One Away* $\frac{1}{8} + \frac{7}{8} \times 57.3\% = 62.6\%$ of the time.

Many games have one potential set of overlapping numbers, allowing elimination of four out of sixteen choices. Consider the initial numbers 28571 for a Kia Sol on September 4, 2019. One should not move the 5 up to 6 *and* the 7 down to 6. Avoiding overlap leads to a 38.5% chance of success on one's second try. With a one in twelve chance of success on the first try, a player would win $\frac{1}{12} + \left(\frac{11}{12} \times 38.5\%\right) = 43.6\%$ of games.

Sometimes there were so many overlapping numbers to be eliminated that winning was essentially a sure bet. On May 7, 2019, a contestant played *One Away* for a Honda Fit LX. The initial number string, 28624, created only five possible choices for nonduplicated numbers for digits 2–5, as shown:

Nonduplicating:	$19,735	$19,715	$19,713	$19,513	$17,513

The remaining eleven combinations have duplicated digits in numbers 2–5, as seen here:

Duplicating	$19,733	$19,535	$19,533	$19,515	$17,735
$17,715	$17,733	$17,713	$17,535	$17,533	$17,515

Believe it or not, the contestant could have ensured himself of victory by guessing $19,735 or $17,513 on his first try and avoiding repeating numbers on his second try. Depending on whether he had one, two, three, or four numbers correct, there would have been only one choice to switch to with no overlapping numbers. Assuming all five possible car prices were equally likely, guessing $19,715, $19,713, or $19,513 on the first try still would have meant an 80% chance of success, 20% on the first try and 75% on the second try for a $P(win) = 20\% + (1 - 20\%) \times 75\% = 80\%$. Unfortunately, the contestant did not win the $19,513 Honda Fit LX. The second guess, $19,715, followed my heuristic, but the first guess of $17,535 duplicated numbers and was effectively wasted.

This Honda Fit LX example is an outlier in there being eleven out of sixteen choices with duplicating digits. Figure 18.11 shows the odds of victory based on the number of duplicate number choices and assuming such choices are avoided.[§]

Options Eliminated with no Overlap	0	4	5	8	10	11	12
Number of Games	4	11	5	7	2	2	1
P(win) with no repeat	31.3%	43.6%	TBD	62.6%	TBD	100%	100%

Figure 18.11 *One Away* Win Probabilities with Elimination of Duplicated Numbers

Excluding ten games in which the contestant picked a car price with duplicated numbers, the win rate would have increased to 36% (8 out of 22) from 25% (8 out of 32). And since miscues were more likely when there were more opportunities for duplication, the ten games excluded would have been easier ones to win without duplication. Specifically, six out of ten such games involved overlap that if avoided would have eliminated eight or more possible combinations.

Overall, I believe *One Away* is winnable about half the time by avoiding duplication, with some setups easier than others. But the showrunners could change this heuristic at any time, which would make the game much more difficult.

Key Tip

Avoid guesses with duplicated numbers in digits 2–5 on one's initial and second guesses.

[§]Based on the three examples shown above. I cannot be sure that all examples with eight options eliminated would lead to the same P(win). For the scenario that eliminates eleven options, 100% win rate assumes choosing one of the two possible first price options that allows for only one nonrepeating second choice. For the scenario that eliminates twelve options, all first price choices with nonrepeating numbers should have only one non-repeating second choice, guaranteeing a 100% chance of victory.

PATHFINDER

Pathfinder is one of the more difficult car games, with a 21% win rate (5 out of 24). Median earnings of $4,250 ranked fifty-third among all pricing games and fifteenth out of sixteen games in the car pricing category. Heuristics and astute guesses for the second digit could have led to modestly improved performance.

The five-by-five *Pathfinder* board looks intimidating, and it is (Figure 18.12). The contestant has four picks for the second digit of the car's price after being given the first digit, three choices for the third digit, two or three choices for the fourth digit (depending on whether one reaches the perimeter after the third pick), and two choices for the fifth digit. It is impossible to have three choices for the fifth digit as the contestant is either on the perimeter or constrained based on his prior path. Eighteen out of twenty-four games had three choices for the fourth pick, whereas the other six times the contestant had reached the perimeter, setting up an up/down or left/right move for the fourth digit. The contestant can win up to three additional tries by selecting the price of small items (priced between $25 and $200) from two possible price choices. Contestants won 69% of additional tries (44 out of 65), which averages to just over two do-overs per contestant.

Figure 18.12 *Pathfinder*. **Illustration courtesy of Omar Faruk**

With three choices for the fourth digit there are $4 \times 3 \times 3 \times 2 = 72$ possible pathways, and with two choices for the fourth digit there are forty-eight possible pathways. This is a large number of pathways given that the contestant gets at most four incorrect choices before losing, depending on whether he picks the correct price for zero, one, two, or three three small items. Fortunately for the contestant, *Pathfinder* is path dependent, so getting each consecutive digit correct, often with the help of do-overs, eliminates many alternative pathways. With three choices for the fourth digit, the probability of winning is 1.4% (1/72) with no do-overs, 6.9% (5/72) with one do-over, 19.4% (14/72) with two do-overs, and a still modest 38.9% (28/72) with three do-overs.

If the chance of winning an additional try is 2/3, then the chance of winning zero do-overs is 1/27, one do-over 6/27, two do-overs 12/27, and three do-overs 8/27 per the binomial distribution. I then multiply probabilities and add to calculate an expected win rate:

$$\left(\tfrac{1}{27} \times \tfrac{1}{72}\right) + \left(\tfrac{6}{27} \times \tfrac{5}{72}\right) + \left(\tfrac{12}{27} \times \tfrac{14}{72}\right) + \left(\tfrac{8}{27} \times \tfrac{28}{72}\right) = 0.1\% + 1.5\% + 8.6\% + 11.5\%$$

$= 21.8\%$; 21.8% is in line with the 21% win rate observed. With two choices for the fourth digit, we just remove twenty-four pathways that corresponded to a third choice for the fourth digit, leaving $4 \times 3 \times 2 \times 2 = 48$ pathways. The odds of winning improve to 1/48 with one pick, 5/48 with two picks, 13/48 with three picks, and 24/48 with four picks. Again, calculate the total probability by multiplying the odds of getting do-overs with the probability of winning conditional on securing those do-overs:

$$\left(\tfrac{1}{27} \times \tfrac{1}{48}\right) + \left(\tfrac{6}{27} \times \tfrac{5}{48}\right) + \left(\tfrac{12}{27} \times \tfrac{13}{48}\right) + \left(\tfrac{8}{27} \times \tfrac{24}{48}\right) = 0.1\% + 2.3\% + 12.0\% + 14.8\%$$

$= 29.2\%$.

If the fourth digit has two choices 25% of the time and three choices 75% of the time, the odds would improve to $(25\% \times 29.2\%) + (75\% \times 21.8\%) = 23.6\%$.

Surprisingly, contestants did no better than random chance on the second digit, picking it correctly six out of twenty-four times on the first try, five out of fifteen times on the second try, and one out of six times on the third try. Failing to outperform random chance on the second digit drove poor results.

Still, we can weight the forty-eight or seventy-two pathways differently to capture opportunities to outperform random chance. Consider the following scenarios:

1. Better odds of success on the second digit, as one would expect the contestant to outperform random chance on the second digit. Specifically, we assume a 40%/30%/20%/10% chance of success on tries 1–4.
2. A 75% chance of success on the fifth digit, as *Pathfinder* is not a game in which the car often ends in a zero or a five. Only once in twenty-four

iterations of the game did the car price end in a zero or five, yet thirteen out of twenty-four choices for the fifth digit were a zero or a five.**

3. Better odds of success on the second digit *and* a 75% chance of success on the fifth digit.

Figure 18.13 summarizes the probabilities of winning: at random, with improved odds for the second digit, with improved odds for the fifth digit, and with improved odds for both the second and the fifth digits. It shows these scenarios assuming three choices for the fourth digit, two choices for the fourth digit, and a 75/25 weighting of the three choice/two choice scenario, as observed on the show.†† I calculate a 38% chance of victory assuming an informed guess on the second digit and avoiding zeros and fives on the fifth digit. This is markedly better than the 22% odds in the random guess scenario and the 21% win rate observed.

	4th Digit 3 Options	4th Digit 2 Options	75/25 Weighting
Random Selection of Car Digits	21.8%	29.2%	23.6%
2nd Digit 40/30/20/10 Odds	29.9%	39.7%	32.4%
5th Digit 75/25 Odds	25.7%	34.1%	27.8%
Improved Odds for 2nd and 5th Digit	34.9%	45.6%	37.6%

Figure 18.13 *Pathfinder* Probabilities Summaries

Are there heuristics to further improve one's chances in *Pathfinder*? Yes. The direction of the fourth digit (i.e., up/down/left/right) was rarely the opposite of the second digit, in fact only once in eighteen games where such a move was possible. For example, if the correct pick on the second digit was left and the third digit up, the fourth digit would be unlikely to be right. This had the effect of reducing the fourth digit choice to two options. The near absence of opposite fourth digit moves would have occurred by random chance in only about 1.5% of sequences of eighteen games.

The second digit of $20,000-plus cars tended be one of the lower number options, with fourteen out of fifteen correct choices corresponding to the lowest or second lowest of the four number choices. This is probably more a function of most $20,000-plus cars on *TPIR* pricing below $23,000 than heuristics per se. I make no adjustments to the analysis underlying the results

---

**The heuristic of avoiding five and zero for the fifth digit was missed by players, who were three for seven in picking the fifth digit correctly. Still, two players had a do-over left for the fifth digit and won on their second try.

††Detailed probabilistic calculations and pathways are discussed at www.popculturemath.com.

in Figure 18.13, as the second digit dynamics are already captured in the improved 40/30/20/10 odds for the second digit.

> ## Key Tip
>
> Do not pick zero or five for the last digit. For $20,000-plus cars, lean toward the lower number options for the second digit unless the car appears more expensive.

POCKET CHANGE

Pocket Change is one of the more original games on *TPIR*. The first part of the game has a pricing component, whereby the contestant guesses the price of the last four digits of the car one digit at a time from among five possible nonrepeating numbers. The second part adds in luck. The contestant is given a default $0.25 and then collects pocket change from four envelopes he has randomly selected after correctly placing each number. His total pocket change needs to be as much as $0.25 plus $0.25 for each miss in pricing the car. Four misses, for example, would require $1.25 of pocket change. The envelopes have pocket change distributed as follows across a total of twenty cards: one $2 card, one $0.75 card, three $0.50 cards, four $0.25 cards, five $0.10 cards, four $0.05 cards, and two nil cards. The expected value across the twenty cards is $0.30.

The most guesses needed to price the car is fourteen (corresponding to ten incorrect guesses), with up to five tries for the second digit, four tries for the third digit, three tries for the fourth digit, and two tries for the fifth digit.

Pocket Change was played only nineteen times with a 37% win rate (7 out of 19). The win rate was in line with other car pricing games, but the median earnings of $7,300 were below the $8,250 category average, as contestants prevailed more when playing for less expensive cars. Contestants won four out of seven times with sub-$20,000 cars, averaging 3.4 misses, and three out of twelve times with $20,000-plus cars, averaging 4.8 misses. Perhaps the showrunners made less expensive cars easier to price.

On June 1, 2020, the contestant needed nine tries, including five incorrect guesses, to guess the price of the $21,086 Hyundai Venue SE. He thus needed $1.50 of pocket change ($0.25 + $0.25 × 5 incorrect guesses) to win. But he had very bad luck, drawing two $0.10 cards followed by two $0.05 cards. His tally of $0.55, including the default $0.25, fell far short of the $1.50 needed to win the car.

The probabilities of winning for each amount of pocket change required are shown in Figure 18.14. If the contestant guesses the price of the car

Car Pricing Guesses	Car Pricing Misses	Pocket Change Required	Probability of Win
4	0	$0.00	100.0%
5	1	$0.25	97.8%
6	2	$0.50	83.7%
7	3	$0.75	65.6%
8	4	$1.00	45.6%
9	5	$1.25	32.5%
10	6	$1.50	24.8%
11	7	$1.75	21.4%
12	8	$2.00	20.3%
13	9	$2.25	17.6%
14	10	$2.50	12.1%

Figure 18.14 *Pocket Change* Win Probability

without any misses, he is guaranteed victory, because he needs $0.25 and gets a default $0.25. The highest amount of pocket change required is $2.50, corresponding to fourteen picks, ten of which are incorrect. The $0.75 or $2.00 card is needed to get to $2.00, whereas only the $2.00 card can get one to $2.50, along with $0.25 across the other three cards and $0.25 of default pocket change. The chance of clearing $2.50 is 12.1%, below the 20% probability of getting the $2.00 card.[‡] The $2.00 card is necessary but not sufficient to clear $2.50. Because money cards do not have evenly spaced values, the probabilities of winning in Figure 18.14 drop off slowly then quickly then slowly as more pocket change is required.

On average contestants missed 4.3 times, needing 8.3 tries. The best pricing result was one miss in five tries on March 27, 2019, although that contestant barely got the $0.50 needed to win a $17,693 Hyundai Elantra after a series of unlucky draws. The worst pricing result was November 7, 2019, when the contestant missed eight of twelve tries in pricing a $20,154 Nissan Kix Crossover. He drew high pocket change cards, ending up with $1.75, but falling $0.50 shy of victory.

Overall, contestants had bad luck in winning pocket change. Given the distribution of guesses to price the car across games, the probabilities in Figure 18.14 would have predicted nine out of nineteen wins (48%). However, chance worked against contestants, who won only seven out of nineteen times (37%).

[‡]To calculate the probability of the $2.00 card, calculate the probability of not getting the $2.00 card in four tries and subtract from 1 as follows: $1 - (19/20 \times 18/19 \times 17/18 \times 16/17) = 1 - 16/20 = 20\%$.

So how many guesses should it take to price the car? If random, it should take nine guesses in expectation:

	Random	Seasons 47 and 48
One, two, three, four, or five guesses for the second digit	3	2.5
One, two, three, or four guesses for the third digit	2.5	2.3
One, two, or three guesses for the fourth digit	2	1.8
One or two guesses for the fifth digit	1.5	1.6
Total	**9**	**8.3**

However, contestants did better on the second digit, averaging 2.5 guesses. There are typically only three reasonable choices for the second digit, meaning that an astute player should average only two choices for the second digit, even better than the 2.5 observed. Astute play might have lowered the average number of observed guesses to 7.8 from 8.3, assuming unchanged outcomes for digits 3–5. And 7.8 average guesses would support a 50%-plus expected win rate. This can be seen by interpolating from Figure 18.14 between the 65.6% expected win rate with seven guesses and the 45.6% expected win rate with eight guesses. One can also explicitly calculate a 52% expected win rate by mapping out the seventy-two different pathways ($3 \times 4 \times 3 \times 2$) to get to the right car price, the number of guesses for each pathway, and the total probability of victory across the various pathways.

Finally, let's evaluate whether there are heuristics that can improve car pricing performance. The number 6 was in the car price in all eleven games in which it was a choice, spread across different car digits. The number 5, on the other hand, was *not* in the car price six out of ten times. Note that these observations have a small sample size. And like all heuristics, there is no guarantee they repeat in future years.[§§]

Key Tip

Plan on using numbers 1 and 6, which were always in the car price during seasons 47 and 48, with 1 often the second digit. Deprioritize the number 5, which was often omitted from the car price.

[§§]The number 6 was always in the car price in Seasons 45, 46, and 49, per www.tpirstats.com, whereas 5 was often *not* in the car—namely, eight out of nine times in season 45, two out of five times in season 46, and four out of nine times in season 49.

STACK THE DECK

With only three wins out of eighteen iterations, *Stack the Deck* was the most difficult car pricing game, which is ironic given the name of the game. Fortunately, it was only played once every twenty episodes. The median car price of $20,700 hardly compensated contestants for the degree of difficulty. It ranked fifty-seventh in median earnings across all pricing games and last among car pricing games.

Consider the game played on January 31, 2019, for a Kia Sol CRV, which was one of three wins. The contestant earned three free numbers, opting for the third, fourth, and fifth numbers, and then proceeded to correctly guess $19,478. Still, the contestant could just as easily have guessed $21,478. If the contestant had opted for free numbers 2, 3, and 4, he would have known he was playing for a $19,47_ car but would not have known if the last number was a 2, 5, or 8. In fact, he might have been less likely to go for 8, thinking it was included as an extra choice for the second digit.

In another win, on December 26, 2019, the contestant played for a Mini Cooper with the numbers 2, 3, 4, 5, 6, 8, and 9, also earning three free numbers. To the audience's surprise, the contestant asked for the first, second, and fourth numbers, at which point he was playing for a $25,_4_ car. Although I understand the contestant wanting the second number, which he picked as his first free number, he should have known without being told that the first number was a 2 (i.e., a $25,000-plus car) and not a 3 (i.e., a $35,000-plus car). Surprisingly, he picked $25,649 from the twelve remaining choices for numbers 3 and 5 and won.

Players were three for three in winning *Stack the Deck* with three free numbers, zero for five with two free numbers, and zero for ten with one free number. No player failed to get at least one free pick. The average number of free picks was 1.6, hardly better than the 1.5 expected if grocery prices were guessed at random.

Estimating the probabilities for *Stack the Deck* is a subjective exercise, as odds are much better than random chance, particularly for the first two digits. Much depends on how one handicaps the odds for the first two digits. If I assume there are two or three equally likely choices for the first two numbers when paired, then I can estimate the player's chance of victory based on the number of combinations from among the free numbers.

- *Zero free numbers*: Beyond the three likely choices for the first two digits, there are five choices for the third digit, four for the fourth digit, and three for the fifth digit, for $3 \times 5 \times 4 \times 3 = 180$ combinations. If one thinks the odds of getting the first two digits correct are 50/50 (because one of three outcomes is more likely), then the odds are 1 in 120, reflecting $2 \times 5 \times 4 \times 3 = 120$ combinations.

- *One free number.* Assume it is used for the third, fourth, or fifth digit. Then there are four possible choices for one of two remaining third, fourth, or fifth digits and three for the last remaining digit. This means $3 \times 4 \times 3 = 36$ combination equivalents given a 33.3% chance on the first two digits (i.e., three likely combinations), and $2 \times 4 \times 3 = 24$ combinations given a 50% chance on the first two digits (i.e., two likely combinations).
- *Two free numbers:* Assume they are used for two of the third, fourth, and fifth digits. Then there are three remaining choices for the other digit. Depending on whether there are two or three likely combinations for the first two numbers, there are $2 \times 3 = 6$ *or* $3 \times 3 = 9$ combination equivalents.
- *Three free numbers:* If the contestant asks for digits 3, 4, and 5, *his odds are between 1/3 and 1/2*—likely closer to 1/2—as a number that shows up in digits 3–5 may be ruled out for the first two digits.

Given the odds above, it should come as no surprise that the contestant was winless with one and two free numbers but won all three times with three free numbers.

Assuming a 60% chance of earning a free pick (versus the low 54% success rate observed), there is a 21.6%/43.2%/28.8%/6.4% chance of 3/2/1/0 free numbers, per the binomial distribution. Odds of winning, even assuming a 50% chance of correctly picking the first two numbers, are still paltry:

$$P(\text{win}) = \left(\frac{1}{2} \times 21.6\% \right) + \left(\frac{1}{6} \times 43.2\% \right) + \left(\frac{1}{24} \times 28.8\% \right) + \left(\frac{1}{120} \times 6.4\% \right)$$

$$= 10.8\% + 7.2\% + 1.2\% + 0.1\% = 19.3\%.$$

Like some other car games, the fifth digit was never zero, despite zero being an option in eight of eighteen iterations of the game. In fact, zero was never the correct third, fourth, or fifth digit. It was the second digit six times and otherwise not in the price of the car. The contestant is well served to consider zero for the second digit but otherwise discard it.[***] The number 5 was absent from the price of the car in season 47, but that reversed in season 48, providing an example of a heuristic unwinding.

[***]Patterns with zero were similar in adjacent seasons, with zero not the fifth digit in the car in six games in season 45, seven games in season 46, and seven games in season 49. And zero was not the third or fourth digit in the car except for two times in season 49.

TEN CHANCES: LISTEN TO THE AUDIENCE

Ten Chances was played more in earlier seasons of *TPIR*. It is a time-consuming game, which likely limits its frequency. But it is an easy game if the contestant follows a time-tested heuristic. Otherwise, it can confuse contestants to the point of losing a very winnable game. Contestants won four out of ten times. Tellingly, most who won did so with chances to spare. Three of the six losses involved contestants not having even one chance to guess the price of the car.

The critical heuristic is that all prices had zero as the last digit. All ten two-digit prizes, all ten three-digit prizes, and all seven car prices revealed ended in zero. Consider the number of combinations that end in zero. There are two combinations ending in zero for the first prize and six for the second prize. For the car, there was never more than one viable first digit, be it a one or a two, which limits the number of realistic car combinations ending in zero to six. These considerations limit the total number of chances required to get all items correct to fourteen—namely, two for the two-digit prize, six for the three-digit prize, and six for the car.

It gets better. For the three-digit prize, in each of the games played, one of the four numbers was clearly too low to be the first number in the prize. For example, on February 26, 2019, the contestant had to price a blender package using the numbers 5, 7, 0, and 2. Clearly the blender package did not begin with 2. The contestant wasted a guess of $275, although he got the blender price, $520, correct on his second try. On October 23, 2019, the contestant was given the numbers 8, 5, 0, and 7 to price a pizza oven/grill. Knowledge of the prize would have suggested 5 was not the first digit. The contestant took six tries to correctly guess the $750 price, initially guessing $580 and subsequently wasting two guesses on price options not ending in zero. Had he eliminated the lowest non-zero number as a possible first digit in the price of the three-digit prize, the number of reasonable choices for that prize would have dropped to four (two choices for the first digit and two for the second digit). Now the game is reduced to twelve choices!

If one is faced with twelve plausible price combinations across the three prizes, the odds of winning are very high. Of the forty-eight possible pathways for the game, only four would require the contestant needing more than ten chances. Odds of winning are thus $\frac{44}{48}$ = 92%. Moreover, thirty-two of the forty-four winning pathways require eight or fewer chances.

Why does *TPIR* make *Ten Chances* so easy to win by having all prizes end in zero? I believe that the showrunners want a decent overall win rate for the game. They know that some contestants will mostly stick to choices ending in zero and win most of the time, whereas some contestants will err by guessing numbers not ending in zero and rarely win. The average is a respectable win rate, similar to the 40% observed. Interestingly, the audience can usually be heard pushing the contestants toward options ending in zero, but not all contestants take notice.

To illustrate the twelve choices in *Ten Chances*, consider the game from May 6, 2019. The contestant's twelve *reasonable* choices across the three prizes were as shown in Figure 18.15.

Garment Steamer (5,4,0)	Nintendo Switch, 7 Games (9,6,0,7)	Chevy Sonic LT (7,9,0,8,1)
Good Choices 50 40	690 670 760 790	17890 17980 18790 18970 19780 19870

Figure 18.15

After a correct $40 guess for the steamer, the contestant tried $670 and $790 (correct) for the Nintendo Switch. After guessing $17,980, $18,970, and $18,790 for the Chevy Sonic, the contestant faltered, guessing $18,079 and $19,087, which did not end in zero. He then guessed $19,780 and $17,089, failing to win the $19,870 Sonic. Given his success on the first two prizes, he could have tried all six reasonable choices for the Sonic, guaranteeing victory.

Key Tip

All guesses for two-digit, three-digit, and car prices should end in zero. Try to eliminate options for the three-digit prize with an unrealistically low first digit. For the car, rule out unlikely second digits.

Listen to the Audience

Ten Chances is one of the few games in which the audience broadly understands the heuristic of all prizes ending in zero. Simply put, the contestant should be listening to the audience. There are many instances in everyday life when our work product can be improved by listening to the audience, as the audience knows the heuristic. In the literary world, literary agents are the audience. They don't have all the answers, but they generally do know what is needed to secure a publishing deal and have a better idea as to what sells with readers. As a writer, do not give up your authenticity, but ignore literary agent feedback at your own peril. Almost every walk of life has its own heuristics or ground rules. Do your best to learn from the audience, who have seen the game played multiple times.

19

Car Pricing Games, Part 2

We now switch our attention to car pricing games in which there is more to the game than just selecting the digits in the car.

CAR DIGITS VERSUS OTHER PRIZES

The contestant guesses digits for the car and for other small prizes.

- *Any Number:* A contestant must use four different numbers from zero through nine to guess digits 2 through 5 in the price of a car. Three of these numbers correspond to the price of a sub-$1,000 prize, and three of these numbers fall into a sub-$10 piggy bank (so many dollars and so many cents). The contestant chooses numbers one at a time, which correspond with the car, the smaller prize, or the piggy bank on the game board. To win, he must pick all four car numbers before picking the three numbers in the smaller prize or the three numbers in the piggy bank, lest he instead win the smaller prize or the piggy bank.
- *Switcheroo:* The contestant is shown a car prize and four small items (sub-$100). He is shown a board with the tens digit missing from each of the small items and from the car prize. The contestant has five blocks corresponding to the missing five digits and a thirty-second time limit to place the blocks in the missing spots. The contestant is then shown how many (but not which) prizes he has correct and is given another thirty seconds to make any changes he would like to make. The contestant then sees whether he wins the car and/or any of the small items.

217

CLOSENESS OF CAR DIGITS

The contestant must get close to the car digits or make over/under decisions.

- *Dice Game*: The contestant is shown the first digit in the price of the car; all digits are numbers one through six. He then rolls a die four times, once for each of numbers two through five in the price of the car. If the first roll does not equal the second digit in the car's price, the contestant guesses whether the second digit in the price of the car is higher or lower. Ditto for the second, third, and fourth rolls. The car price is then revealed one digit at a time to see if all numbers match the player's selections (i.e., higher or lower).
- *Lucky Seven*: The contestant is given seven $1 bills to start the game and is shown the first digit in the price of the car. He must then guess the remaining digits one at a time. After each digit is guessed, the actual digit is revealed. The contestant pays the difference between his guess and the actual digit in dollars, owing nothing if he chooses the exact number. For example, a guess of 5 when the digit is 7 costs $2. If the contestant pays out all his dollars, the game ends. If the contestant has at least $1 remaining after the last digit is revealed, he buys the car for $1 and wins the game.

CLOSE PRICE GUESS

The contestant must pick a price close to the price of the car.

- *Card Game*: A contestant draws from a deck of cards to arrive at a price that is sufficiently close to the price of the car without going over. He first draws a "reference card" from a special deck of seven cards (two $1,000 cards, two $2,000 cards, two $3,000 cards, and one $5,000 card) to see how close he must be to the price of the car without going over. The game then shifts to a full deck of cards with a card's value equal to the card number times $100 (with face cards valued at $1,000). The contestant can make an ace any value. The contestant starts at $15,000 and draws until he believes he is sufficiently close to the price of the car without going over (i.e., less than the reference card amount) or until he uses an ace to jump to a target price. The car price is then revealed to see if the contestant wins.
- *Five Price Tags*: There are five price tags, one of which is the price of the car. The player is shown four small items (usually sub-$200) and must decide for each one whether the displayed price is true or false. If he

answers correctly, he wins the small item and gets to choose one price tag. He can win up to four small items and four price tag selections. If one of his price tag selections is the price of the car, he wins the car.

- *That's Too Much*: The contestant is shown ten five-digit prices in increasing sequential order, left to right. One at a time, he tells Drew whether to keep going or whether the price is too much relative to the price of the car, by saying "that's too much." The contestant wins if he selects the first price tag in excess of the car price.

ANY NUMBER

Any Number is a classic *The Price Is Right* (*TPIR*) car game. The 29% win rate (12 out of 42) and $6,450 median earnings were below average for car pricing games and ranked twenty-fifth overall. In the thirty losses, the contestant priced the smaller three-digit prize seventeen times and the piggy bank thirteen times. The game in Figure 19.1 aired on June 4, 2019. Here the contestant won a

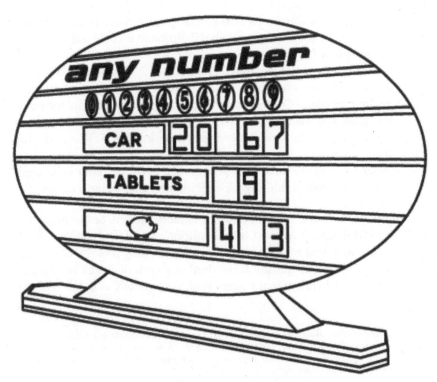

Figure 19.1 *Any Number. Illustration courtesy of Omar Faruk*

$598 pair of tablets, missing out on the $20,167 Kia Forte but avoiding the $4.23 piggy bank. He reached a point with one missing digit in each row, and then selected 5, the first digit in the tablets.

Though there is not strategy per se in *Any Number*, there are notable heuristics to improve one's chances.

- *Zero is an unlikely digit in the car, except for the second digit.* Although it is common for zero to be the last digit in the price of the car in some pricing games, this is not the case for *Any Number*. In forty-two iterations of the game, zero was the fifth digit only twice, the third digit once, and the fourth digit once. It did appear as the second digit (i.e., a $20,XXX car) on six separate occasions. Zero was the last number in the price of the three-digit prize seventeen times, and fifteen times it was the second digit in the price of the piggy bank. Notably, five "last pick" zeros resulted in losses.

- *Start with the second digit in the car.* Sorting out the second digit in the car first allows one to avoid guessing zero for the third, fourth, or fifth digits, which is fraught with risk. Contestants followed this strategy, as twenty-two of forty-two times their first correct pick in the car was the second digit. Most of the other times contestants still seemed to target the car's second digit first.

- *Number 2 was hardly ever the third digit in the car.* If down to the third digit in the car, do not select the number 2, although it could still show up as the fourth or fifth digit in the price of the car.

- *Number 2 was piggy bank danger.* The number 2 showed up twenty-one out of forty-one times in the piggy bank. (Once the piggy bank price was not revealed.) Contestants did avoid the number 2 toward the end of the game, as it was chosen only twice as the final digit in the piggy bank.

- *The first digit in the three-digit prize was never less than 5.* If only the first digit is remaining in the three-digit prize, choose from numbers 1–4 to avoid winning the three-digit prize in lieu of the car. Four losses could have been avoided this way (albeit without guaranteeing a car win).

Are these observed heuristics due to random chance, or are they more likely by *TPIR* design? A Monte Carlo simulation suggests that random chance alone would rarely lead to the observed heuristics—only 3 to 4% of the time would 0 show up so infrequently as the third or fourth digit, and less than 5% of the time would 2 end up in the piggy bank as frequently as it did.* The data

---

*The simulations were run with nine numbers instead of ten based on the assumption that one number is likely for the second digit of the car and therefore not available for digits 3–5 of the car, for any digit in the three-digit prize, or for any of the three digits in the piggy bank. This is a stricter setup than running a simulation with all ten digits.

is less convincing in terms of ruling out 0 as a fifth digit, as the least represented fifth digit still showed up twice in 15% of simulations.

Contestants seemed to avoid the numbers 0 and 2, choosing them twenty-two times and twenty-one times, respectively, versus twenty-eight times for the average number. Moreover, many times that 0 and 2 were chosen, it was early and to target the second digit in the car.

So how much better can one do with these heuristics? Since my heuristics were initially based on season 47, I tested them in season 48 when one number remained for each of the car, the three-digit prize, and the piggy bank. Contestants went winless in these five games, whereas my heuristics would have led to three wins based on process of elimination and random picks from the remaining options as necessary. On this basis, season 48 wins would have been increased from 20% (4 out of 20) to 35%, a meaningful boost. Consider the June 4, 2019, game in Figure 19.1. The contestant's last pick was between the third digit in the price of the car, the first digit in the tablets, and the second digit in the piggy bank. His number choices were 1, 2, and 5. Using my heuristics, (1) number 5 was the likely first digit in the tablets (because three-digit prizes are always >$500) and (2) number 2 was likely in the piggy bank (number 2 being piggy bank danger), implying number 1 was likely the third digit in the car, which it was.

Key Tip

Tackle the second car digit first, avoiding 0 as a number choice once you resolve the second car digit. See heuristics discussion for numbers likely in the piggy bank or the first digit of the three-digit prize. Apply heuristics later in the game.

SWITCHEROO

Switcheroo is a funny game, both because of the music played and because of contestants' obsession with pricing the small items, often after deciding on the digit for the car. After *Stack the Deck*, *Switcheroo* tied with *Pathfinder* as the second most difficult car pricing game, with only four wins out of nineteen tries and $4,350 median earnings. The odds of winning the car on one's first try are hardly better than a one in five random chance. Furthermore, the contestant gains only marginal information after his first guess with which to better his odds on the second guess.

To the extent pricing knowledge is of value in *Switcheroo*, it involves correctly guessing the prices for all the sub-$100 products, such that the remaining block is the fourth digit in the car price. This is easier said than done. In the June 20, 2019, episode captured by Figure 19.2, the contestant guessed $21,365

Figure 19.2 *Switcheroo. Illustration courtesy of Omar Faruk*

for the car on his first go and decided to stick with his initial picks after seeing that he had three prizes correct in his initial selection. He won the car, a Mazda CX3 Sport. He also correctly guessed $90 for the four cutting boards and $39 for an oil diffuser, while reversing numbers for an $82 laminator and a $57 hairdryer. Although he ruled out the 9 and the 3 for the car with his small item picks, he clearly had less visibility on the middle numbers and could have just as easily placed the number 6 with the laminator or the hairdryer.

Heuristics can provide a meaningful benefit in *Switcheroo*. Consider the following observations:

1. The number 5 was never the fourth digit in the price of the car, despite being an option in seventeen of nineteen iterations of the game.

2. The fourth digit in the car duplicated the second, third, or fifth digit in the car in three games but only once when there was but one choice for a duplicated number. In the other eleven games with one duplicate number option for the fourth digit, the contestant would have been well served to avoid that choice.

Turning to probabilities, consider a random guess on the first try. On average,

- *five numbers are correct* with probability $\frac{1}{120}$ = 0.8% (5 × 4 × 3 × 2 = 120 combinations, one with all right);
- *three numbers are correct* with probability $\frac{10}{120}$ = 8.3% (120 combinations, ten with three right);
- *two numbers are correct* with probability $\frac{20}{120}$ = 16.7% (120 combinations, twenty with two right);
- *one number is correct* with probability $\frac{45}{120}$ = 37.5% (120 combinations, forty-five with one right); and
- *zero numbers are correct* with probability $\frac{44}{120}$ = 36.7% (120 combinations, forty-four with zero right).

If the player gets five digits correct on his first try, he wins. If two or three numbers are correct, he wins the car with probability three out of five or two out of five, so he is not well served to change his initial picks. If he has one number correct, he is indifferent between the one in five chance of winning with his existing pick or the one in five chance of winning with a changed pick. If he has zero numbers correct, he gets a second guess at the car with one in four odds. Odds of winning are thus:

$$\left(\frac{1}{120} \times 1\right) + \left(\frac{10}{120} \times \frac{3}{5}\right) + \left(\frac{20}{120} \times \frac{2}{5}\right) + \left(\frac{45}{120} \times \frac{1}{5}\right) + \left(\frac{44}{120} \times \frac{1}{4}\right)$$
$$= 0.8\% + 5.0\% + 6.7\% + 7.5\% + 9.2\% = 29.2\%.$$

The 21% win rate slightly underperformed the above odds, which assumes a random first guess and sensible decision making as to whether to change one's car pick on the second guess. On the contestant's first attempt, eight out of nineteen tries ended with zero correct (42%), seven ended with one correct (37%), two ended with two correct (10.5%), and two ended with three correct (10.5%). Contestants mostly behaved sensibly, with underperformance due to chance. When the contestant got three correct, he correctly kept his car guesses, both times winning the car. When he had two correct, he incorrectly changed his pick once, although he would have lost either way. The two other wins came on second guesses after initially having zero right.

Now let's add a heuristic and assume that one of the five options can be eliminated because it is a five or because it repeats the second, third, or fifth digit in the car price. If I avoid that choice and assume the heuristic holds, my odds improve versus random chance as follows:

- *five numbers correct*: $4 \times 4 \times 3 \times 2 = 96$ combinations, one with all five numbers correct;
- *three numbers correct*: There are nine combinations when ruling out one number choice for the car, and six still win. The probability of winning the car is 6/9, or 2/3;
- *two numbers correct*: There are only seventeen combinations, as three would have the avoided number being used in the car price. The odds of winning with two numbers are therefore 8/17;
- *one number correct*: There are only thirty-six combinations, as nine would have the avoided number being used in the car price. On the second pick the odds of winning are 1/4, as there are four valid choices;
- *zero correct*: The remaining thirty-three of ninety-six combinations. There is a 1/3 chance of winning the car on a second try (as one can rule out the eliminated number and the number initially selected).

Odds of winning are now:

$$\left(\frac{1}{96} \times 1\right) + \left(\frac{3}{32} \times \frac{2}{3}\right) + \left(\frac{17}{96} \times \frac{8}{17}\right) + \left(\frac{3}{8} \times \frac{1}{4}\right) + \left(\frac{11}{32} \times \frac{1}{3}\right) = 1.0\%$$
$$+ 6.3\% + 8.3\% + 9.4\% + 11.5\% = 36.5\%.$$

Voilà! We have again used an observed heuristic to meaningfully increase *Switcheroo* odds from 29.2% to 36.5%, at least for those games in which one number can be excluded.

But is my heuristic-based assumption reasonable? For the most part. I believe that the number 5 is a very unlikely fourth digit in the car *and* the fourth car digit is very unlikely to repeat the second, third, or fifth digit. Twelve of nineteen games had one repeating car number choice and seventeen games had five as an option. Ten games were eligible for both heuristics, although in three of those ten games, 5 was the repeating number choice, which prevented one from making use of both heuristics. In seven games the contestant could apply both heuristics and eliminate two choices.

> ## Key Tip
>
> In choosing the fourth digit of the car, avoid the number 5 and number choices that repeat the second, third, or fifth digit of the car. Do not change your car guess if you have two or three numbers correct.

DICE GAME

Dice Game had a high 48% win rate (16 out of 33) and $10,350 median earnings, ranking seventh among all pricing games and third among car pricing games. Because *Dice Game* requires all car digits to fall between 1 and 6, the car must be a $20,000-plus car (19 cases) or a sub-$17,000 car (14 cases), as occurred on June 11, 2019, and illustrated in Figure 19.3.

Figure 19.3 *Dice Game. Illustration courtesy of Omar Faruk*

In *Dice Game*, rolling a 1 or 6 is a godsend and rolling a 3 or 4 makes for a hard call on the part of the contestant. The contestant missed the fifth digit only twice in thirty-three games but missed the second digit three times, the third digit seven times, and the fourth digit six times. Only once did he miss two digits in seventeen losses.

The strategy in *Dice Game* is probability sprinkled with heuristics. If one rolls a 2 or a 5, and it is not a direct match, then it is wise to go higher with a 2 and lower with a 5. The only exception is the second digit if one believes the car is worth $16,000–$16,666 or $21,000–$21,666. Ten out of fourteen cars starting with a 1 were $16,000–$16,666 cars, whereas ten out of nineteen cars starting with a 2 were $21,000–$21,666 cars.

What should the player do if he rolls a 3 or 4 and it is not an exact match for the number in the car? For the second digit, his decision should clearly be based on what he thinks the price of the car is. For digits 3–5, consider the distribution of numbers in seasons 47 and 48, as shown in Figure 19.4.

Number	3rd Digit	4th Digit	5th Digit
1	3	3	1
2	3	11	5
3	7	3	5
4	6	7	5
5	4	6	17
6	10	3	0
Total	33	33	33

Roll 3	3rd Digit	4th Digit	5th Digit
1-2 (lower)	6	14	6
4-6 (higher)	20	16	22

Roll 4	3rd Digit	4th Digit	5th Digit
1-3 (lower)	13	17	11
5-6 (higher)	14	9	17

Figure 19.4 Car Digit Frequency—*Dice Game*

A contestant *rolling a 3* (when not an exact match) would have been well served to go higher for the third number, as the third number was a 4, 5, or 6 in twenty contests and a 1 or 2 in six contests. Similar for the fifth number, when a 4, 5, or 6 showed up in twenty-two contests versus a 1 or 2 in six contests. For the fourth number, it was a toss-up, as numbers 4, 5, or 6 and 1 or 2 showed up with equal frequency.

A contestant *rolling a 4* (when not an exact match) would have fared similarly whether going higher or lower on the third digit. On the fourth digit, going lower would have been advantageous, given a 1, 2, or 3 in seventeen contests and a 5 or 6 in nine contests. On the fifth digit, the contestant should have gone higher. Although the car never ended in a 6, seventeen times it ended in a 5, versus eleven instances of ending in a 1, 2, or 3.

Contestants generally followed the above heuristics on the third and fifth digits. They erred in choosing higher three out of five times when rolling a 4 for the fourth digit. All three times they were wrong and lost.

Consider the probability of success in *Dice Game*. Assume to start that the digits in the car are random numbers between 1 and 6. When the contestant rolls a die, there is a one in six chance he has the number correct. Rolls of 1 or 6 ensure success. Rolling 2 means guessing higher and being correct four-fifths of the time, since there are four higher numbers and one lower number. And vice versa when rolling a 5. Rolls of 2 or 5 lead to correct guesses with probability of $\frac{1}{6} + \left(\frac{5}{6} \times \frac{4}{5}\right) = \frac{1}{6} + \frac{2}{3} = \frac{5}{6}$. Rolling 3 or 4 should cause one to guess higher or lower, respectively, correctly three-fifths of the time, since there are three digits in one direction and two in the other. So a roll of 3 or 5 is successful with probability $\frac{1}{6} + \left(\frac{5}{6} \times \frac{3}{5}\right) = \frac{1}{6} + \frac{1}{2} = \frac{2}{3}$. Averaging 100% success with rolls of 1 and 6, 5/6 success with rolls of 2 and 5, and 2/3 success with rolls of 3 and 4 means a 5/6 chance of success on each number, and a $\left(\frac{5}{6}\right)^4 = 48\%$ chance overall.

Now consider that there is greater certainty on the second digit, where players were correct 91% of the time (30 out of 33). Using $\frac{11}{12}$ for success on the second digit instead of 5/6 would imply a 53% chance of success, based on $\left(\frac{5}{6}\right)^3 \times \frac{11}{12} = 53\%$. Offsetting this ease on the second digit is greater difficulty on the fourth digit, as well as the fifth digit for contestants not aware that *Dice Game* cars commonly end in 5. Specifically, rolling a 3 for the fourth digit creates more of a coin toss as to whether the number is higher or lower, whereas rolling a 4 for the fifth digit will cause a contestant to guess lower if erroneously assuming randomness. If I assume 50% odds of success when rolling a 3 for the fourth number and 40% odds of success when rolling a 4 for the fifth number (i.e., incorrectly assuming randomness), the odds of victory in *Dice Game* drop back down to 50%. Conversely, if I choose to go higher when I roll a 4 on the fifth number, odds edge back up to 52%.

Key Tip

For the most part, after dice rolls, pick higher or lower based on which rolls offer more possible numbers. Exceptions would be the second digit, often a 6 if the car starts with 1, and often a 1 if the car starts with 2. The fifth digit is also more likely higher when rolling a 4, as half the cars ended in 5.

LUCKY SEVEN: CONDITIONAL STRATEGIES

The most played car game, *Lucky Seven* was played fifty-five times during seasons 47 and 48, with a 31% win rate (17 out of 55) for $6,650 in median earnings, ranking twenty-third across all pricing games. *Lucky Seven* (Figure 19.5) is one of the more exciting pricing games in terms of crowd excitement and contestant emotion. During season 47, in a likely first for *TPIR*, a contes-

Figure 19.5 *Lucky Seven. Illustration courtesy of Omar Faruk*

tant guessed each digit of the car correctly without losing one dollar. This occurred during Dream Car Week on May 31, 2019, when the contestant priced a $24,635 Nissan Frontier king cab perfectly. The next best playing of *Lucky Seven* during seasons 47 and 48 had the contestant losing $3.

The perfect game aside, *Lucky Seven* is daunting, although strategy and heuristics can be of some benefit. Strategy is tied to conditional decision making—how the contestant should vary his remaining guesses based on how many dollars he has lost on his prior picks. At the end of the game, there is only winning and losing. To analogize to football, if time is running out and a team's offense is driving at midfield, it needs to go for a Hail Mary on its last play versus a twenty-yard throw short of the end zone. There is no prize for losing by a little.

Conditional Strategies

Hail Marys are present in business and finance but are usually discouraged for good reason. Traders who have put up sizable losses at investment banks have a history of hiding those losses and taking sizable bets to try and return their book to a profitable position, at great risk to the bank. The most memorable example is that of derivatives trader Nick Leeson bringing down Barings Bank in 1995. His initial speculative trades were very profitable, earning the bank 10 million (mm) British pounds (GBP), or 10% of its annual profit in

1992. Leeson was given the responsibility of checking his own trades, which allowed him to hide his losses and take huge risks with the firm's capital. Losses of 23mm GBP in 1993 mushroomed into 208mm GBP in 1994, which precipitated a series of risky bets on the value of the Nikkei stock index in early 1995. These trades faltered, and Leeson's total losses amounted to 827mm GBP, twice the bank's trading capital. Barings went bankrupt in February 1995. Indeed, debtholders usually have strict provisions limiting actions of equity holders when a firm is near bankruptcy. The last thing debtholders want is assets being sold to briefly delay bankruptcy while jeopardizing their recovery in bankruptcy.

Let's start with second-digit dynamics, move on to heuristics, and then discuss conditional strategies. Most second digits are a 7, 8, or 9 if playing for a sub-$20,000 car or a 1, 2, or 3 if playing for a $20,000-plus car, although nicer models can run upward of $24,000. The contestant should be cautious about picking a 1 (as in a $21,000–$21,999 car) or a 9 (as in a $19,000–$19,999 car) unless he feels very certain that one of those numbers is the second digit in the car. A $24,000–$24,999 or a $16,000–$16,999 car can leave the contestant with only four dollars left after the first guess, whereas guessing a 2 or 8 allows the contestant to modestly miss in either direction. During seasons 47 and 48, four cars were priced between $24,000 and $24,999, and three between $16,000 and $16,999, which would have put the contestant in a deep hole with a 1 or 9 pick. One caveat regarding the second digit: if the contestant strongly feels that the car is a sub-$18,000 car, he is better off selecting a 7 as the second digit and, similarly, for selecting 3 if he believes the car is a $23,000-plus car.

Turning to heuristics, consider which numbers tend to be closest to the third, fourth, and fifth car digits in *Lucky Seven*. Is the contestant better off guessing a 5 if he has the leeway to play it safe on the later numbers, or should he guess a 4 or a 6? Figure 19.6 addresses this question. It shows average dollars

$ Lost	2nd Number Pick		3rd Number Pick			4th Number Pick				5th Number Pick						
	1/9	2/8	4	5	6	4	5	6	7	3	4	5	6	7	8	9
0	25	12	7	5	8	6	2	2	6	9	2	6	2	1	8	15
1	12	35	13	15	9	10	8	8	10	12	15	4	7	10	16	8
2	10	7	14	12	20	11	14	14	16	6	12	10	10	21	2	1
3	7	0	4	19	12	6	17	22	6	2	1	18	24	2	6	2
4	0	0	13	4	6	8	14	9	8	1	8	15	10	9	2	6
5	0	0	4	0	0	14	0	0	9	8	15	0	0	10	9	2
6	0	0	0	0	0	0	0	0	0	15	0	0	0	0	10	9
7	0	0	0	0	0	0	0	0	0	0	0	0	0	0	0	10
Total	54	54	55	55	55	55	55	55	55	53	53	53	53	53	53	53
Avg	$0.98	$0.91	$2.27	$2.04	$1.98	$2.76	$2.60	$2.51	$2.49	$3.09	$2.81	$2.60	$2.62	$2.72	$2.85	$3.28

Figure 19.6 *Lucky Seven*—**Simulation of Different Digits Picked and Dollars Lost**

lost and the distribution of dollars lost across different picks for the third digit (numbers 4–6), the fourth digit (numbers 4–7), and the fifth digit (numbers 3–9). It also compares picking a 1 or a 2 as the second digit for a $20,000-plus car or an 8 or 9 as the second digit for a sub-$20,000 car.

For the second digit, picking a 2 or 8 would have led to average losses of $0.91 and never a $3 loss. On the third digit, the contestant would have done okay by picking 5, but could have done slightly better by picking 6. Not only was the average $1.98 lost with a third-digit "6 pick" better than the average $2.04 lost with a "5 pick," but there also would have been fewer big misses. A "6 pick" would have lost the contestant $3 or $4 only eighteen times versus twenty-three such losses with a "5 pick." On the fourth digit, there was a greater skew toward higher numbers, with a pick of 6 being slightly closer on average than a pick of 5. Picking a 7 would have even performed a touch better. There is the risk of some game-upending $5 deviations with a 7 pick, but overall there would have been fewer $3, $4, or $5 misses.

On the fifth digit, conditionality is in play, so Figure 19.6 is less helpful if the contestant has $1 or $2 left. For the fifth digit, it is better to look at the number of cars ending in each digit, as follows:

Last Car Digit	1	2	3	4	5	6	7	8	9
Number of Games	0	10	9	2	6	2	1	8	15

If one picked a 2 or an 8, followed by picks of 6, 7, and 6 for the third, fourth, and fifth digits respectively, he would have won only eight out of fifty-five times (15%). This is actually worse than observed performance.

If one instead makes the fifth-digit pick conditional on the number of dollars left, performance materially improves. Based on the frequency of numbers in the last digit of the car, consider the following strategy for the fifth digit conditional on the number of dollars remaining:

- with $1 left (i.e., cannot lose even $1), pick 9, as it is the most frequent number;
- with $2 left (i.e., can lose $1), pick 8, resulting in twenty-three wins, or 3, resulting in twenty-one wins;
- with $3 left (i.e., can lose $2), pick 7, resulting in thirty-two wins, or 4, resulting in twenty-nine wins;
- with $4 left (i.e., can lose $3), pick 6, resulting in forty-three wins; and
- with $5 left (i.e., can lose $4), pick 5. You can never lose.

If a contestant has $2 or $3 left for the fifth digit, he will have to choose whether to go with 8 and 7 respectively (i.e., higher conditional choices) or 3 or 4 respectively (i.e., lower conditional choices). Be aware that 8 and 9 were far more prevalent numbers in season 48 (15 out of 26 games) than in season 47 (8 out of 27 games).

We can test this strategy for the fifth digit. Doing so suggests twenty-one wins in fifty-three games with higher conditional choices and twenty wins with lower conditional choices. If we simplify the above to simply picking the number 5 in the event that the contestant has $4 or more left, the win tally improves to twenty-three with higher conditional choices and twenty-two with lower conditional choices. This is more than a 40% win rate. Albeit a counterintuitive result, it has to do with the number 9 being less likely in seasons 47 and 48 when the contestant reached the fifth number with $4 or more left. I recommend a $5 guess with $4 or more remaining, which should be nearly as successful and far simpler.

Key Tip

Choose 2 or 8 for the second digit for a $20,000-plus or sub-$20,000 car. Go with 6 for the third digit, 7 for the fourth digit, and choose the fifth digit conditional on how many dollars remain: pick 9 with $1 remaining, 3 or 8 with $2 remaining, 4 or 7 with $3 remaining, and 5 with $4 or more remaining.

CARD GAME

Card Game is a game of patience and luck. Are you lucky enough to avoid selecting the $1,000 reference card, which requires you to be within $1,000 of the price of the car without going over? And are you patient enough to wait for the tally to climb above $20,000, given that the median car value was nearly $23,000 (which was higher than other car pricing games) and never less than $21,000? The low 35% win rate in *Card Game* (11 out of 31) suggests that contestants either were not patient enough or were prone to underbidding.

Both factors were likely at play. Contestants likely underestimated the price of the car, particularly given a richer car mix. And this dynamic was exacerbated in *Card Game* because it takes a large number of draws to get from $15,000 to more than $20,000. The average value in a deck of cards, excluding aces, is $617 (four face cards at $1,000 and the 2–9 cards at $200–$900). On average, eight non-ace draws are required to get to $20,000, whereas ten such draws are required to get to the $21,000–$22,000 range, a good stopping point for a $23,000 car.

In seventeen of the twenty losses in *Card Game*, the contestant was under the price of the car by more than the reference card. On average, when losing, the contestant stopped $3,200 below the price of the car versus $1,800 below the price of the car when winning. Never once did a contestant draw more than eight cards.

Aces are not hard to come by. A contestant has a 50% chance of landing an ace by his eighth draw, whereas ten non-ace draws on average are required to reach $21,000. Hence an ace will be drawn more than half of the time before a contestant stops, assuming he does not stop too early. The formula for the probability of getting an ace after n draws is $1 - \left(\frac{48}{52} \times \frac{47}{51} \times \ldots \times \left(\frac{48-n+1}{52-n+1} \right) \right)$. In seasons 47 and 48, players drew aces in fifteen out of thirty-one games, with twelve of the fifteen aces drawn before the tally exceeded $19,000.

Contestants fared only slightly better when drawing an ace, winning seven out of fifteen times versus four out of sixteen without an ace. But this is not an apples-to-apples comparison. Eight of fifteen aces were drawn when the player had a $1,000 reference card, making victory more difficult. Otherwise, performance would have likely been better, as the average stopping value was $1,350 below the car price with an ace versus $4,000 without. Only five of the eight losses with the ace involved guesses too low, which speaks to bids being less biased downward.

A big determinant of whether contestants won, given a proclivity to stop too soon, was the price of the car. Contestants won nine of eighteen times when the price of the car was between $21,000 and $23,000 and two of thirteen times when the price of the car was more than $23,000. In sixteen out of thirty-one iterations of the game, contestants erred by stopping below $20,000. They actually won five times when stopping below $20,000, because in four of those five games they had a reference card of $3,000 or $5,000.

Only two of the eight losses with the ace involved stopping below $20,000. Contestants bid more aggressively with the ace, and they needed to, given that ace draws tended to occur in games with a $1,000 reference card. Underbidding was more pronounced when the contestant had to make a large number of draws from the deck. Impatience as a driver of underbidding paralleled what was observed in *Range Game* and *That's Too Much*.

All in all, *Card Game* performance was helped by a healthy number of ace draws and hampered by a higher-than-expected number of $1,000 reference cards. Contestants would have performed much better if they played patiently. A simple heuristic would be as follows:

- On an ace draw, choose a car price of at least $21,000.
- Assuming no ace draw, continue drawing until reaching $20,100, given that the lowest priced car was $21,100. Draw once more upon crossing the $20,000 mark if playing with a $1,000 reference card.

- If the car seems more expensive, like a crossover SUV, add $1,500 to the formula above. If the car is a semi-premium car (like a Mini Cooper or a Jeep), add $3,000 to the above.

I have no doubt that the win rate of *Card Game* would easily exceed 50%, and probably approach 75%, using the above strategy. This assumes car prices remain in historical ranges.

Key Tip

With an ace, guess at least $21,000. Without an ace, draw until reaching at least $20,100, or draw once more with a $1,000 reference card. Add $1,500 to the above if the car seems more expensive, and $3,000 if a semi-premium car.

FIVE PRICE TAGS

One of the more straightforward car games, *Five Price Tags* was played only nineteen times during seasons 47 and 48. The 53% win rate and $11,000 median earnings ranked first among car pricing games. During Dream Car Week, contestants played for a $67,000 Range Rover and $80,000 Lincoln Navigator. Although the Lincoln was won, this did not materially affect the median values.

To win up to four chances in *Five Price Tags*, the contestant must decide if the displayed price for a small item (sub-$200) is true or false. On average, contestants won 2.7 tries. Contestants earned three tries in fourteen games, two tries in four games, and only one try in one game. Of course, more tries at the car means a greater chance of victory, with contestants winning eight of fourteen times with three tries, one of four times with two tries, and the one time with only one try. Consider the odds of success if the contestant has slightly better than random odds of victory, with probabilities of success on the first through fifth tries of 25%, 22.5%, 20%, 17.5%, and 15%. Two tries here would suggest a 47.5% chance of success and three tries a 67.5% chance of success. Given the distribution of tries earned, these probabilities would suggest a 60% chance of success, slightly higher than observed.

In general, there was a $4,000 range from the low price tag to the high price tag, with the price tags approximately $1,000 apart. Lower price tags were often the correct picks, with the lowest and second-lowest price tags each the right choice seven times in nineteen games. The first, second, and third highest price tags were the right choices only once, three times, and once, respectively. This skew is unlikely due to chance. In a Monte Carlo simulation of nineteen contests with each price tag 20% likely, only 3% of simulations resulted in two price tags accounting for at least fourteen of nineteen wins. This is strong evidence that *TPIR* set up the games to favor the lowest and second-lowest price tags. Contestants should be able to prevail more than 60% of the time if they bias their guesses toward the two lowest price tags versus the second- and fourth-lowest price tags that they favored in seasons 47 and 48.

Contestants could also have done much better at winning tries at the car. Their picks went counter to the game's heuristics. Fifty-seven out of seventy-six guesses for the small items were "false," yet "true" was the correct pick for fifty-one out of seventy-six items. Such a high skew of "true" outcomes would occur less than 0.5% of the time by chance in seventy-six trials per the binomial distribution. Moreover, items listed with a price of $100 or higher were "false" only two out of nineteen times, whereas contestants guessed "false" sixteen out of nineteen times for such items.

Key Tip

"True" is more likely for small items, and especially so if the displayed price is more than $100. Lean toward the lowest choices in one's car picks unless confident in a more expensive car.

THAT'S TOO MUCH: PATIENCE REVISITED

That's Too Much is a very difficult car game. The 22% win rate (11 out of 49) and $22,850 median prize value led to $5,250 in median earnings, ranking thirty-eighth among pricing games and thirteenth among car pricing games.

One can readily see why the game is so difficult: ten price cards and one correct pick. The odds of winning would be one in ten if the contestant had no insight on the price of the car and no heuristic insights about when to stop. Fortunately, there are heuristics at play. For starters, the two lowest price cards and the two highest price cards were never correct, and the eighth card was correct only once. The contestant seemed to play accordingly, never stopping after the first or second revealed price and never staying in the game long enough to see the eighth, ninth, or tenth stopping price. That makes the game a bit easier, more like a one-in-five guess.

Next, consider the distribution of outcomes versus the distribution of guesses, as shown in Figure 19.7. Twenty-three of thirty-seven losses involved stopping one price card too early or one price card too late. This suggests a player is deciding from among three prices about two-thirds of the time and choosing the correct price from among those three options about one-third of the time, with a $\frac{2}{3} \times \frac{1}{3} = 22\%$ chance of success. Of course, this matches the observed win rate.

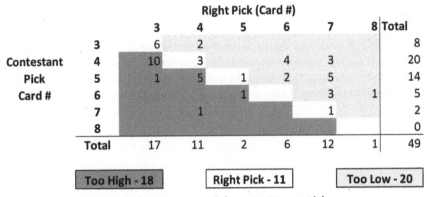

Figure 19.7 *That's Too Much*—**Contestant Pick versus Correct Pick**

The losses were split fairly equally between stopping too early and stopping too late. However, the early stops were often far too early. Ten times the contestant stopped two cards too early, and three times he stopped three cards too early. Only once did he stop two cards too late, and only once three cards too late. This suggests a bias to stop too soon, like what was observed in *Card Game* and *Range Game*. On average, the contestant stopped after 4.45 cards, whereas on average, the right play was to stop after 4.75 cards. Nine of the eleven wins occurred when the contestant stopped before the fifth card, including six wins on the third card and three wins on the fourth card.

Fortunately, heuristics can improve one's odds of winning *That's Too Much*. Although the contestant may be tempted to stop on the fifth card, the fifth card was the winning pick only twice, so best to avoid it. In contrast, the third and seventh cards were the winning stopping points seventeen and twelve times respectively (and near 60% of the time combined).

So how exactly should the contestant proceed? First, gauge whether he thinks the car price is in the lower or upper half of the price cards. If in the lower half, lean toward the third card. If in the upper half, lean toward the seventh card. Although one does not know the price of the later cards before they are revealed, the delta between two adjacent price cards typically ran between $1,200 and $1,500. After the third card is revealed, the player can anticipate a seventh card that runs $5,000 to $7,500 higher in price than the third card. He can then ask himself whether a $5,000 to $7,500 higher price tag is a reasonable later stopping point or not.

I suspect that these heuristics would have improved odds of winning to better than 40%.[†] Why? If contestants simply flipped a coin and stopped after the third or seventh price card, they would have won about 30% of the time. If they had been able to pick between the third and seventh cards with two-thirds accuracy, the win rate would have increased to $\approx 2/3 \times 3/5 = 2/5$, or 40%.

[†]The third and seventh cards were overrepresented in seasons 45–48 but were right only six out of twenty-one times in season 49.

Key Tip

Avoid the fifth price tag. Prioritize the third or seventh price tag depending on whether you think the right answer is in the lower half of the price tags or the upper half of the price tags, with adjacent cards $1,200 to $1,500 apart. Flex up from the third to the fourth tag or stop on the sixth tag as appropriate.

## 20

Car Plus Games

In car plus games, the objective is to price the car, but the contestant can also win other prizes worth more than $1,000 without winning the car. In some cases, the other prizes themselves can lead to a $5,000-plus outcome, which I define as a win, even if the contestant does not win the car. I also include in car plus games two unusually difficult games: *Three Strikes*, in which a contestant plays for a luxury car, and *Triple Play*, in which a contestant plays for three generic cars. Most car plus games are played infrequently, with *Golden Road*, *Three Strikes*, and *Triple Play* played only a handful of times per season.

Car plus games are a challenging category of games, as seen in Figure 20.1. The 30% category win rate fell short of 36% for car pricing games. In many wins the contestant won only the non-car prizes. In *Golden Road*, five of seven contestants won all the non-car prizes, but only one of them won the luxury car at the end of the "golden road." And in none of the five *Temptation* "wins" did the contestant win the car. Consequently, median earnings of $3,950 in car plus games fell far short of the $7,500 observed in car pricing games. In effect, the higher prize value was more than offset by lower odds of winning the game's main prize. Not one car plus game had median earnings of more than $8,000.

Game	Played	Wins	Losses	Win Rate	Average Prize	Median Prize	Avg Win	Median Win	Min Win	Max Win	Median Earnings	Earnings Rank
Gas Money	20	4	16	20%	$30,346	$30,358	$22,513	$26,781	$ 9,000	$27,490	$ 5,356	36
Golden Road*	7	5	2	71%	$86,638	$80,863	$20,116	$ 7,721	$ 6,731	$70,383	N/A	N/A
More or Less	18	5	13	28%	$18,470	$24,142	$13,681	$ 7,259	$ 6,354	$24,611	$ 4,094	56
Pass the Buck	22	9	13	41%	$22,262	$20,468	$13,240	$16,195	$ 5,000	$23,115	$ 7,443	16
Temptation	15	5	10	33%	$27,527	$26,523	$ 6,571	$ 5,360	$ 4,940	$11,470	$ 2,666	71
Three Strikes*	8	0	8	0%	$56,857	$57,993	N/A	N/A	N/A	N/A	N/A	N/A
Triple Play*	7	1	6	14%	$56,503	$57,694	$54,435	$54,435	$54,435	$54,435	N/A	N/A
Total/Wtd Avg	97	29	68	30%	$34,009	$34,265	$16,051	$14,104			$ 3,965	

Note: In *Pass the Buck*, the contestant has one to three chances to win the car, but can also win $1,000, $3,000, or $5,000 in a pre-car pick. I consequently add $3,000 to each of the prize values as a representative "middle-of-the-road" outcome.
* Not used to calculate or rank median earnings

Figure 20.1 Statistical Summary—Car Plus Games

In terms of prize value, there is no richer game than *Golden Road*, with a median prize value of about $80,000, reflecting the ultra-luxury car prizes, including a Porsche 911, a BMW530i, and a Mercedes E400. *Three Strikes* and *Triple Play* were next, each with a $58,000 median prize value, *Triple Play* reflecting three cars at nearly $20,000 apiece. In *Gas Money, More or Less*, and *Temptation*, one can win the car only by winning all non-car prizes—$10,000 in *Gas Money* and about $5,000 in *More or Less* and *Temptation*. In *Pass the Buck*, one can win cash totaling $1,000 to $8,000 prior to winning the car.

Here I provide a brief synopsis on how each game is played.

- *Gas Money*: A contestant is shown five possible price tags for the car. Here the contestant must choose the four *incorrect* price tags one by one, leaving the actual car price tag for last. Each of the four incorrect price tags has a cash prize of $1,000, $2,000, $3,000, or $4,000, and the contestant can stop at any point with his cumulative cash winnings. If he picks the actual car price tag in any of his first four picks, he loses the car and his running cash total.

- *Golden Road*: A progressive game, a player must correctly price a three-digit prize and a four-digit prize for a chance to win a luxury car, usually valued at about $70,000. The player keeps the earlier prizes if he falters at a later stage of the game. For the three-digit prize, the player is shown two possible prices with a different first digit (hundreds digit) and must select the correct price from the two options. For the four-digit prize, the player is shown three possible prices with a different second digit (hundreds digit) and selects the correct price from the three options. Finally, for the five-digit car, the player is shown four possible prices with a different third digit (also hundreds digit) and selects the correct price from the four options. Not only are there more choices as the game progresses, but pricing knowledge is of less benefit in choosing the correct price.

- *More or Less*: A progressive game, the contestant is shown an incorrect price for each of four prizes, the last pertaining to the car. He must guess if the real price is more or less. If correct, he plays for the next prize, of greater value, until reaching and potentially winning the car. The first prize tends to average $1,000 in value, the second and third prizes slightly more than $2,000, and the car about $20,000.

- *Pass the Buck*: There are six cards, numbered one through six, one with $1,000, one with $3,000, one with $5,000, one with a picture of a car, and two with "lose everything." The contestant has one free pick. He is then shown two sets of two grocery items with corresponding prices. In each set of two, one price is correct, and one needs a dollar added to it. The contestant gains an extra pick by selecting the grocery prize that needs a dollar added to it. With his one, two, or three picks, the contestant then chooses from the six cards. After each pick, the contestant can stop with his cumulative prize money or continue to play for the car. A "lose everything" card resets things at zero, although the player can continue to play if he has remaining picks.

- *Temptation*: The contestant is shown, one by one, four prizes whose total value is typically between $4,000 and $7,000. Each prize uses only two different digits in its three- or four-digit price. He must select which of the two digits in the first prize is the second digit in the price of the car, and so forth for the second, third, and fourth prizes. After a chance to change any numbers, the contestant must decide whether to play for the car and the prizes or simply to take the four smaller prizes.

- *Three Strikes*: The five numbers in the price of a car are shown and then placed into a bag with three strikes, all of which are circular tiles. If the contestant draws a strike, the strike is removed. Upon drawing a number, he must guess the corresponding digit in the price of the car. If correct, the number is removed from the bag. If incorrect, the number is placed back in the bag. The player wins if he correctly places all car numbers before drawing three strikes.

- *Triple Play*: Without going over, the contestant must guess which of two displayed prices is closest to the price of the first car, which of three displayed prices is closest to the price of the second car, and which of four displayed prices is closest to the price of the third car. If the contestant gets all three correct, he wins all three cars; otherwise, he wins nothing.

GAS MONEY

Gas Money is as hard, if not harder, than it appears. Four out of twenty wins and median earnings of $5,350 underscore the challenge. Three wins were for the full $10,000 and the car, whereas one win had the contestant walking with $9,000 in cash versus playing for the final $1,000 and the car.

To see why contestants should never quit while they are ahead, consider the expected value of winning versus stopping with cash. With an average $20,000 car—and assuming the contestant values the car as such—he would need to have $15,000 in cash with two choices left, $10,000 with three choices left, or $7,500 with four choices left to justify stopping. All are impossibilities. The most cash a contestant can have banked is $9,000 with two choices left, $7,000 with three choices left, and $4,000 with four choices left. Even a less expensive $18,000 car valued at a 20% discount (i.e., $14,400) would not justify stopping.[*]

---

[*]For an $18,000 car valued at a 20% discount, the expected value of playing on, including the $10,000 cash, is $1/3 \times (18,000 \times 0.8 + \$10,000) = \$8,130$. But $7,000 is the maximum bankable cash after two guesses.

Only once did the contestant stop with cash. He had $9,000 with one pick remaining versus what turned out to be a $16,310 car. Assuming 50% odds of victory, only a 50% discount on that $16,310 car would have justified stopping: $(1/2 \times 16{,}310 + \$10{,}000) \times 1/2 = \$9{,}155$. In every other iteration of *Gas Money*, the contestant rightly kept playing, including twice when he had amassed $7,000 cash after two picks.

The heuristics for *Gas Money* are very helpful, although backward looking. Similar to *Five Price Tags*, the car prices were disproportionately the lower choices. In twenty games, the car price was the highest price twice, the second-highest price only once, never the middle price, eleven times the second-lowest price, and six times the lowest price. The overrepresentation of the two lower price tags (17 out of 20) in *Gas Money* was even more pronounced than in *Five Price Tags* (14 out of 19). Using binomial probabilities, the odds of any two price tags being the price of the car seventeen out of twenty times by way of random chance is less than 0.1%. And the odds of one price tag being the correct pick more than half the time in twenty trials are about 0.3%.

The contestant could have won more than half the time by never choosing the second-lowest price, and he could have eliminated the middle price with an early pick every game. However, many contestants thought the middle price might be the car, never picking it first in twenty tries, picking it second only once in seventeen tries, and never picking it third in nine tries. Instead, contestants chose the highest and lowest numbers as likely *non-car* prices with the greatest frequency, and the second highest and second lowest choices with modest frequency, as seen in Figure 20.2.

Price Rank	1st Pick		2nd Pick		3rd Pick		4th Pick		Car Price Category	Car Losses	Car Wins
	Tries	Losses	Tries	Losses	Tries	Losses	Tries	Losses			
Most Expensive	6		4	1	1	1	1		2	2	0
2nd Highest	2		2		3		0		1	0	1
Middle	0		1		0		2		0	0	0
2nd Lowest	1	1	6	5	2	2	0		11	8	2
Least Expensive	11	2	4	2	3	2	0		6	6	0
Total*	20	3	17	8	9	5	3	0	20	16	3

* Contestant stopped with $9,000 after 3rd pick on 10/4/18

Figure 20.2 *Gas Money*: Picks and Car Price by Price Rank (Number of Observations)

Contestants lost three times on their first pick, eight times on their second pick, and six times on their third pick. Three car wins in nineteen tries with one stoppage for the cash prize suggests that contestants did not exhibit any edge in ruling out the wrong car price options in *Gas Money*. Avoiding the lower price tags would have provided a nice advantage.

Key Tip

During seasons 47 and 48, the middle price tag was rarely the car. The fourth-highest ranked price tag often is. Only strong risk aversion justifies stopping early for cash, even with $8,000 or $9,000 in cash.

GOLDEN ROAD

Golden Road is not that golden and was played only seven times across two seasons. Contestants won five times, four times by winning more than $5,000 in prizes but not the car. Only one contestant won the luxury car grand prize, taking home a $62,000 BMW 530i on December 17, 2018, along with $8,000 in kitchen appliances and a $700 cookware set.

Not once did a contestant fail to win the first prize. The two possible choices typically differ in price by at least $400, sometimes $600 or $700. For example, on February 18, 2020, the contestant had to choose between $460 and $960 for an electric mountain bike; $960 was obviously the correct price.

The second prize (i.e., the four-digit prize) is more of a jump ball, although pricing knowledge can help. Here the three prices differ in their hundreds digit. On the December 23, 2019, episode, the player had to choose from among $5,961, $5,761, and $5,061 for a pizza oven. He correctly guessed $5,761 and got to play for the grand prize, a Jaguar I-PACE EV 400, which he did not win. The second prize was won an impressive five out of seven times. This was a function of the middle price being the correct choice six out of seven times, and the contestant choosing the middle price four times.

As for the car, good luck. Four price choices with a different hundreds digit means the contestant can bring little price knowledge to bear. Although a negligible sample size, one win in five tries is essentially a random guess. Interestingly, the correct car price was the lowest priced option four times, including the one car win. Whether this proves to be forward looking or backward looking is anyone's guess.

Key Tip

Prioritize the middle price option for the second prize. Favor the lowest price option for the car.

MORE OR LESS

More or Less is a progressive game like *Golden Road*, albeit with four prizes and only two options for each prize: "more" or "less" than the displayed price. Contestants won five out of eighteen times, twice winning the car and three times winning only the first three prizes but surpassing $5,000 in prize value.

The first prize is set up to be won easily, and it was won all eighteen times. When the prize was less than the displayed price, it was on average half the $1,690 price shown. On June 2, 2020, a $739 digital smoker was half of the $1,500 displayed price. When the prize value was more than the displayed price, it was on average $1,118, nearly twice the $606 price shown. Never was the correct pick "more" when the displayed price was more than $1,200, nor was the pick "less" when the displayed price was less than $1,200.

After the first prize, the choices become more difficult. Still, contestants were correct fifteen out of eighteen times for the second prize. "More" and "less" were again each the right answer about half the time. The delta between the actual price and the shown price continued to average about $600 but relative to prizes that averaged $2,000. Had a contestant always guessed "more" when shown a price of $2,000 or less and always guessed "less" when shown a price more than $2,000, he still would have erred three times, including on November 16, 2018, when a virtual reality package had a displayed price of $1,500 but only cost $540.

The third prize was again more challenging than the second, with nine wins in fifteen tries. Here the median differential between the actual and displayed price was only $400. Notably, "more" was the correct answer *five out of eight times* with higher shown prices (i.e., more than $2,000), and "less" was the correct answer *seven out of eight times* with lower shown prices (i.e., $2,000 or less).

Finally, the car: nine players reached this stage, but only two prevailed. Only once was the difference between the car price and the shown price greater than $1,000, making the "more" or "less" decision essentially a random guess. There were seven "more" guesses for the car, but "less" was the correct answer seven out of nine times. This was likely a function of chance given the limited data points.

Key Tip

For the first and second prizes, "more" tends to be correct for low prices and "less" for high prices, with $1,200 and $2,000 respective cutoff points. For the third prize, high prices may be more likely to be "more" and vice versa, so proceed with care. The car is a coin toss but "less" was more often the correct answer.

PASS THE BUCK

Pass the Buck was played twenty-two times during seasons 47 and 48. Contestants won the car five times and earned more than $5,000 in cash on four other occasions. Median earnings of $7,450 ranked sixteenth, in part because of small cash earnings in many iterations of the game. Players rarely walk away empty-handed.

Similar to *Gas Money*, the contestant always maximizes his expected earnings by continuing to play for the car rather than walking away with a cash prize. Figure 20.3 shows that with an $18,000 car (the average during seasons 47 and 48, excluding one special show), there is never a scenario whereby the contestant should choose the cash versus continuing to play for the car. This holds true even if applying a 10% discount to the value of the car, whether due to risk aversion or a lower perceived value. Always playing for the car may seem counterintuitive; it is a function not just of the car value but of the high cash value often won along with the car.

Total Picks	Picks Left	Cash Prize In Hand	Expected Value Further Play
2	1	$1,000	$5,800
2	1	$3,000	$6,600
2	1	$5,000	$7,400
3	1	$1,000	$7,250
3	1	$3,000	$8,250
3	1	$5,000	$9,250
3	1	$4,000	$7,750
3	1	$6,000	$8,250
3	1	$8,000	$8,750
3	2	$1,000	$9,600
3	2	$3,000	$10,000

Figure 20.3 *Pass the Buck*: Expected Value of Further Play

Only twice did a contestant quit with cash, once after drawing $8,000 on the first two of three draws and once after drawing $5,000 on the first of two draws. The expected value of playing on through the last pick would have been $8,750 in the first instance and $7,400 in the second instance, evidence of some risk aversion.

Figure 20.3 shows the expected win value with each of one, two, and three picks for each decision point in the game. With one pick there is no decision to make.[†]

We can now calculate an expected value for the game assuming the contestant wins an extra pick 60% of the time, in line with the 61% observed. The contestant won two additional picks, for a total of three picks, ten times out of twenty-two, one additional pick seven times, and no additional picks five times.

- *One pick* should occur 40% × 40% = 16% of the time, with expected earnings of $4,500, averaging the six cards (0, 0, $1,000, $3,000, $5,000, and $20,000 car).
- *Two picks* should occur 2 × 60% × 40% = 48% of the time with expected earnings of $6,900, based on averaging 6 × 5 = 30 possible combinations of cards.
- *Three picks* should occur 60% × 60% = 36% of the time with expected earnings of $11,200, based on averaging 6 × 5 × 4 = 120 possible combinations of cards.

Based on the frequency of one, two, and three picks, expected earnings for *Pass the Buck* are $8,064, modestly higher than the $7,450 observed, which included two early stops.

> ## Key Tip
> Never take the cash. You are better off on average playing for the car plus cash.

TEMPTATION: A BIRD IN HAND

In *Pass the Buck*, the contestant is never well served in expectation by taking the cash in lieu of playing for the car (or more cash). In *Temptation*, the reverse is true. The contestant should always take the prizes. Even if the contestant is 100% certain of the correct second digit of the car, his odds are still not much

---

[†]Consider two total picks with one pick remaining and a cash card of $1,000. There is a 1/5 chance of $4,000 = $1,000 + $3,000, a 1/5 chance of $6,000 = $1,000 + $5,000, a 1/5 chance of $19,000 = $1,000 + $18,000 car, and a 2/5 chance of $0. Consider three total picks with one pick remaining and a cash card of $1,000. There is a 1/4 chance of $4,000 = $1,000 + $3,000, a 1/4 chance of $6,000 = $1,000 + $5,000, a 1/4 chance of $19,000 = $1,000 + $18,000 car, and a 1/4 chance of $0. Consider three total picks with one pick remaining and two cash cards of $4,000. There is a 1/4 chance of $9,000 = $4,000 + $5,000, a 1/4 chance of $22,000 = $18,000 + $4,000, and a 2/4 chance of $0. For three total picks with two picks left, there are twenty scenarios to average.

better than one in eight of winning the car along with the four prizes. But he can walk away with four prizes whose median value was $5,350. The expected value of winning a $20,000 car and $5,350 of prizes only 1/8 of the time is $\frac{1}{8} \times \$25,350 = \$3,170$.

Temptation is really that simple. Yet in fifteen contests, players walked away with the prizes only eight times. Had they taken the prizes all fifteen times, median earnings would have been $5,350, twice the $2,650 observed.

In none of the fifteen iterations of *Temptation* did the contestant price the car correctly, although in the December 20, 2018, episode, the contestant had the car price correct at $43,915 before changing it to $43,912. I would have expected one or two correct car prices out of fifteen iterations of the game, regardless of whether the contestant played for the car. However, contestants picked the second digit correctly on only six out of fifteen tries. Surprisingly, contestants did better on the third, fourth, and fifth digits, guessing them correctly eleven, twelve, and nine times respectively. Presumably there would have been at least one correct car price guess had contestants had better success on the second digit.

The odds of winning the car in *Temptation* are made somewhat easier with a simple heuristic—in all fifteen iterations of the game the car price ended in 5 or 0, and twelve of those iterations involved a prize in which the contestant chose from the digits 5 or 0 and another number. The other three episodes had the contestant choosing between 5 and 0 for the fifth digit, with 5 the correct number all three times.

Still, this heuristic, even if guaranteeing 100% accuracy on the fifth digit, does not increase the odds enough to justify playing for the car and the prizes. One is still left with odds not much better than one in eight, given difficulties on the second digit. If one had been offered $5,000 in prizes, he would need one in five odds to justify playing for a $20,000 car and the $5,000 in prizes, and higher with risk aversion. This would require near certainty on both the fifth digit and the second digit. So enjoy *not* playing *Temptation*. Do not get carried away trying for the car.

Key Tip

Take the prizes rather than playing for the car, unless you (a) have near certainty on the second digit, (b) have the opportunity to choose 5 or 0 for the fifth digit, and (c) are not risk averse.

A bird in hand is worth two in the bush is one of the most overused phrases. Sometimes it is in one's best interest to take one's initial job offer, sometimes it is not. More often than not we take our good fortune as a sign of broad "buyer interest," because we believe in ourselves or what we are trying to sell. When selling a house in a balanced housing market, it is very perilous to risk one formal offer in order to court prospective interest from other buyers. We waited nine months to get an initial offer when we sold our house in 2017. At the time there were other parties circling and the interested buyers were tough negotiators. Still, they wanted the house, and we focused on getting to yes, which proved to be in our best interests, as interest from the other parties faded. Indeed, in a balanced housing market or job market, there should on average be one buyer for every seller. A bird in hand is more representative of how things should play out on average in these markets, just like in *Temptation*.

THREE STRIKES

Three Strikes may be the hardest pricing game. It was not won once during seasons 47 and 48 across eight contests. Although contestants may get excited about the chance to win a luxury car worth more than $50,000, their chances of winning are de minimis. Fortunately for contestants, *Three Strikes* is rarely played.

The odds of winning are both simple and complicated. If the contestant knew the car price a priori, he would still have only a 3/8 chance of winning. To see this, rethink the problem as one of drawing a strike on one's last draw. Since there are five numbers and three strikes, the odds of winning are identical to a strike being the last tile in the bag, namely 3/8. An alternative way to see this is to realize that there are twenty-one combinations of numbers and strikes that allow for victory: one with no strikes, five with one strike, and fifteen with two strikes, depending on when the numbers are drawn versus when the strikes are drawn. Each of the twenty-one paths to victory has a 1/56 chance, for a total probability of 21/56, or 3/8. We can extend this analysis to six or more tries.

What is unique about *Three Strikes* is that it combines sampling with replacement (when a number is guessed incorrectly and placed back in the bag) and sampling without replacement (when the contestant gets a strike or guesses a number correctly and it is removed from the bag). As a result, whether a contestant guesses a number incorrectly early or later alters his odds. For example, if the contestant guesses a number incorrectly on his first draw, he has back-to-back draws, each with a 5/8 probability of pulling a number. If he guesses the number incorrectly on the fourth number draw, he has back-to-back draws, each with a 2/5 probability of pulling a number, since three numbers would already be on the board.

Calculating the odds requires that I make an assumption about the likelihood of that error being on the contestant's first, second, third, or fourth number draw. For simplicity, I equally weight each winning scenario and calculate a win probability of 26.2%, much lower than the 37.5% chance with no wrong guesses.[‡] Under the same assumption, seven or eight tries (i.e., two or three numbers incorrect) leads to modeled odds of 18.6% and 13.0%, respectively.

I believe—given *informed* play—that eight tries is a realistic expectation for the quantity of number draws required to price the car. This suggests a win probability of 13% assuming an equal likelihood of missing numbers earlier versus later. This would mean completing roughly ten draws before drawing the third strike, as there are far more winning paths with two strikes (28) than there are with one strike (7) or zero strikes (1). In two games out of eight, the contestant made it to eight number draws before drawing the third strike. But these two contestants were not "informed," with one incorrectly guessing six numbers and the other incorrectly guessing seven numbers.

What makes for informed play? First, a contestant chooses from the highest three numbers for the first digit, as the first digit was never one of the lowest two numbers during seasons 47 and 48. Second, the contestant chooses only one of the lowest two numbers for the fifth digit, as the fifth digit was never one of the highest three numbers during seasons 47 and 48.

Three reasonable number choices for the first digit and two reasonable number choices for the fifth digit narrows the number of choices from an unrestricted 120 choices ($5 \times 4 \times 3 \times 2 \times 1$) to a restricted 36 choices $3 \times (3 \times 2 \times 1) \times 2$. Specifically, there are three choices for the first digit and two for the fifth digit, leaving three choices for one of the second, third, or fourth digits, two choices for the second of the second, third, or fourth digit, with the remaining second, third, or fourth digit a plug. However, the number of potential incorrect guesses is far fewer, because the number of incorrect guesses is additive. At most, the contestant can *incorrectly guess* the first digit twice, the fifth digit once, one of the second, third, or fourth digits twice, and the second of the second, third, or fourth digits once. So at most, six possible incorrect guesses and eleven total number draws. On average, the contestant will incorrectly guess three numbers, as guessing the first digit correctly on one's first try is about as likely as guessing it incorrectly once or guessing it incorrectly twice. Same for the fifth digit across two options and so forth.

The reduction in the number of choices as described earlier requires that the contestant get the first digit and fifth digit in place first, given that there

---

[‡]My bias was to weight earlier misses as more likely, as more digits in the car price are open. However, and somewhat surprisingly, contestants had better success in correctly placing the first number drawn.

are a limited number of options for each. Consider initially drawing one of the lowest two numbers—it has a 50/50 chance of being the last digit in the price, and a 1/3 chance of being placed correctly in one of the three middle digits otherwise. If you guess it as the fifth digit and are wrong, you now know the fifth digit with certainty and have reduced the possibilities for the misplaced number from four to three choices. In contrast, guessing the low number as a middle digit and being wrong does not provide as much useful information. The same goes for drawing one of the three highest numbers. Guessing the first digit provides the most information in terms of limiting the number of remaining possibilities, whether correct or not.

Consider the episode on December 18, 2019, played for a Lincoln Aviator Reserve SUV priced with numbers 0, 1, 4, 5, 6 (Figure 20.4). The contestant drew his third strike on his tenth draw, having incorrectly guessed seven numbers before that point. Because he did not tackle the first and fifth digits first and because he guessed a high number for the fifth digit, he still had two possible incorrect number choices even after seven incorrect guesses.

Number Draw	1	2	3	4	5	6	7	8	9	10
#/Pick	5/2nd	6/2nd	Strike	4/1st	1/4th	5/1st	5/5th	Strike	4/2nd	Strike
Correct	N	N		N	N	N	N		N	

Figure 20.4

The correct price for the Lincoln Aviator was $60,451. Had the contestant targeted the first and fifth digits first, he would have incorrectly guessed 5 as the first digit and then correctly guessed 6 as the first digit. At that point there would have been one less number in the bag, making our revisionist exercise less relevant. But had the next number drawn been a 4, he should have guessed the second, third, or fourth digit, and had the following number been a 1, he should have guessed the fifth digit. By not focusing on the first and fifth digits, the contestant missed opportunities to limit his options.

If the contestant is not aware of the heuristic for the fifth digit, there would be four choices for the fifth digit. The possible number choices, again additive, would be $3 + 4 + 3 + 2 + 1 = 13$ and up to eight possible incorrect guesses. The player on average would need nine number picks to price the car (i.e., four incorrect picks), which lowers the win probability further to 9%.

Key Tip

Choose one of the three highest numbers for the first digit. Choose one of the lowest two numbers for the fifth digit. Do not target the second, third, or fourth digit before resolving the first and fifth digits.

TRIPLE PLAY

Triple Play is fun to watch but not as fun to play. It was played seven times in two seasons and was won only once.

The Price Is Right (*TPIR*) makes the first car an easy win. Not only were contestants six for seven on the first car, but the more expensive option was the correct pick all seven times. The first car was priced at $18,000 on average, compared to $15,600 for the average higher priced option and $13,500 for the lower priced option. A car priced less than $15,600 is virtually unheard of at 2019–2020 price levels, making the lower priced option a nonstarter.

On the second car, the game gets trickier, and contestants were two for six. For the second car, the correct selection was the low price on four occasions and the middle price on two occasions, whereas the contestant chose the middle price five times and the highest price once. The show from April 4, 2019, had the contestant guessing the middle price, $21,460, for a $20,259 Toyota Corolla when the correct choice was $19,250, the lowest price. *TPIR* likely avoids making the high price the correct choice for the second car because it makes the high price the correct choice for the first car. Only two contestants made it to the third car. Once the contestant guessed the highest price and the right choice was the second-highest price. The other time he correctly selected the third-highest price.

A few observations from *Triple Play*, some of which border on heuristics:

- Car price choices generally remain equally spaced as one progresses through the game. The delta between each option and the next higher or lower option was about $2,100 for the first car, $2,800 for the second car, and $2,000 for the third car.
- The second car was never less expensive than the first car, and the third car was never less expensive than the second car. A similar relationship held for the correct price picks (i.e., the closest price without going over). The two lowest priced options could have been eliminated for the third car in one of two games on this basis.

Putting these observations together, *Triple Play* need not be such a difficult game. The first car should be won every time. Past play suggests that the second car should be the lower or middle priced choice. The third car is probably not the cheapest option, given that the third car was always more expensive than the second car, and the second car was always more expensive than the first car. With two realistic choices for the second car and three for the third car, odds should be no worse than one in six. A one-in-six chance for three cars worth $55,000 would imply $9,000 in expected earnings, better than most car games.

Key Tip

Guess the highest price option for the first car and avoid the highest price option for the second car. Each car price and winning pick is likely more expensive than the prior one.

## 21

Other Car Games

The final category of car games includes the remaining five in which the contestant neither has to price the car nor has the potential for an outsized win. It includes three of my favorite games, *Let 'Em Roll*, *Rat Race*, and *Spelling Bee*. Although no common element links these games, many have high win rates, as seen in Figure 21.1. *Hole in One (or Two)* ranked highest among all games in median earnings at nearly $12,000 and featured a 69% win rate. *Rat Race* had the fifth highest median earnings at $10,950 with a 55% win rate, aided by favorable chance.

Game	Played	Wins	Losses	Win Rate	Average Prize	Median Prize	Avg Win	Median Win	Min Win	Max Win	Median Earnings	Earnings Rank
Hole in One	16	11	5	69%	$23,203	$17,716	$17,808	$17,315	$13,185	$23,840	$11,904	1
Let 'em Roll	23	7	16	30%	$25,154	$18,985	$38,596	$17,070	$5,000	$100,000	$6,891	22
Master Key	16	4	12	25%	$23,216	$23,198	$23,389	$23,389	$21,924	$25,106	$6,522	25
Rat Race	20	11	9	55%	$21,676	$21,791	$19,321	$19,169	$16,614	$23,346	$10,964	5
Spelling Bee	19	9	10	47%	$21,989	$18,162	$17,955	$18,141	$15,251	$20,475	$9,407	9
Total/Wtd Avg	94	42	52	45%	$23,112	$19,917	$22,232	$18,515			$9,057	

Figure 21.1 Statistical Summary—Other Car Games

Each of these other car games was played with moderate frequency (i.e., 15–25 times). Though contestants do not need to know the price of the car in the other car games, they do need to know the prices of groceries or small items. Such knowledge allows them a closer putt in *Hole in One* or more tries in *Let 'Em Roll*, *Master Key*, *Rat Race*, and *Spelling Bee*. Luck of course also plays a major part in deciding whether they win the car.

- *Hole in One (or Two)*: The contestant tries to price rank six grocery items in order from least expensive to most expensive. The more items placed in ascending order, the closer to the hole he putts. If he gets all six items priced correctly in ascending order, he putts from the sixth line, only one or two feet from the hole, and wins a $500 bonus. He gets two chances to putt the ball in the hole to win the car, from whichever line he putts from.

- *Let 'Em Roll*: There are five cubes, each of which has a car on three faces and money on the other three faces, typically $500, $1,000, and $1,500. The contestant starts off with one roll. He can win up to two more rolls by guessing whether a second grocery item is more or less expensive than a first grocery item, whose price is shown, and by guessing whether the third grocery item is more or less expensive than the second (after the second item's price is revealed). With one, two, or three rolls, the contestant walks up a set of stairs and rolls the three dice down a ramp. If all five cubes show a car, he wins the car. Otherwise, he can keep the money shown or re-roll the cubes with money (but not re-rolling the car cubes) to try for five cars on his next roll. If he has not won the car by his final roll, he keeps the sum of the money from his final roll.

- *Master Key*: There are five keys and five key slots, one of which opens the car prize, one of which opens a second prize worth about $2,000, and one of which opens a third prize worth about $1,000. The fourth key is a master key that opens all three prize key slots. The fifth is a blank and opens zero prize key slots. The contestant starts with zero keys but can win up to two keys. He is shown a sub-$100 item and a three-digit number, with the price either the first two or last two digits of the three-digit number. If he prices one or two items correctly, he selects one or two keys from the rack and tries the locks to see what he wins.

- *Rat Race*: Five different colored wind-up rats run a race on a track. The contestant wins rats by guessing the price of three smaller items within a certain margin of error. To earn a rat(s), he must estimate the price of the first item, a sub-$10 item, within $1; the price of the second item, a sub-$100 item, within $10; and the price of the third item, a $100-plus item, within $100. He then picks the colors of his rats, and the rat race begins. If a contestant's rat wins first place, he wins the car; second place, the ~$2,000 prize; and third place, the ~$1,000 prize.

- *Spelling Bee*: There is a board with thirty numbered hexagonal cards. On the backs of the cards, unseen, are eleven *C*'s, eleven *A*'s, six *R*'s, and two car icons. To win, the contestant must pick cards that spell "car" or display one of the two car cards. He starts off with two picks and can win up to three more picks by guessing the price of three sub-$100 prizes within $10. After having picked his numbers, the backs of the cards are revealed one at a time. Each card is worth $1,000. The contestant can stop at any time and claim cash in the amount of $1,000 for each unrevealed card. If, after revealing the cards, the contestant spells "car" or holds a car card, he wins the car. If not, the contestant goes home empty-handed.

HOLE IN ONE (OR TWO): SUFFICIENT FOR SUCCESS

Back in the day, *Hole in One* offered contestants only one putt for an expensive car. Contestants won infrequently. Starting in 1986, *Hole in One* became *Hole in One (or Two)*. Contestants stand a good chance of getting the ball in the hole in one or two putts, particularly if they are putting closer to line six than to line one. They need not be avid golf players to accomplish this feat. Contestants won eleven out of sixteen times (69%), even as they putted from the first line (i.e., the farthest putt) in seven of sixteen iterations of the game (Figure 21.2). Remarkably, they won five out of seven times putting from the first line, twice on the first putt, and three times on the second putt.

Putt Line	1	2	3	4	5	6	Total
Attempts	7	3	3	1	1	1	16
Wins	5	1	3	1	0	1	11
Win 1st Putt	2		2	1		1	6
Win 2nd Putt	3	1	1				5

Figure 21.2 Outcomes in *Hole in One (or Two)*

One might think there is no strategy in *Hole in One (or Two)*. But contestants can approach the ordering of grocery items more strategically in order to putt closer to the hole (as illustrated in Figure 21.3). They did rather poorly in ordering the grocery items, limiting their chance of a closer putt. On average, they

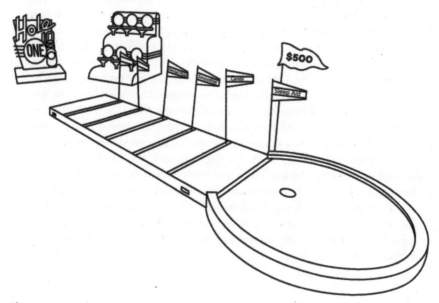

Figure 21.3 *Hole in One (or Two)*. Illustration courtesy of Omar Faruk

ordered the first 2.3 items correctly, a poor result given that random guesses would have led to a 1.9 average. Contestants would have been well served by trying to place three or four out of six items in sequential order and overlooking items for which they were uncertain about price rather than going for six in a row and making a mistake early on. Putting from the third line is not too challenging of a setup, especially with two opportunities to putt the ball in the hole.

Key Tip

Try to place three or four groceries in ascending price order to putt from line three or four versus trying to place all six grocery items in ascending order and erring early in the process.

Sufficient for Success

The ease of *Hole in One (or Two)* masks contestant shortcomings. In ranking groceries, contestants are better off determining what is sufficient for success—namely, getting three or four items in ascending order so they can putt closer to the hole. Overreaching for all six items can lead to failure. Now consider the battle for leadership in the immuno-oncology market, in which drugs stimulate one's immune system to fight cancer. Merck's Keytruda and Bristol Myers Squibb's (BMY) Opdivo were undergoing simultaneous clinical trials starting in 2015 to treat solid state tumors, notably non–small cell lung cancer (NSCLC). BMY had the lead, with early approved indications for Opdivo in cancers including melanoma and renal cell carcinoma. Opdivo 2016 sales of $3.8 billion far exceeded Keytruda's $1.4 billion. But first-line NSCLC was the big opportunity. First-line treatments are the first treatment given for a disease, often part of a standard set of treatments. BMY went big in its first-line trial, targeting a broad swath of NSCLC patients. Merck on the other hand went narrow, limiting its trial to individuals pretested for significant expression of the PD-L1 biomarker, a protein that in the lab suggested better responsiveness to immuno-oncology drugs. Merck's trial succeeded with its narrowly defined population, whereas BMY's trial failed with its broad population, setting up Merck's Keytruda to dominate not just NSCLC but other indications. In 2020, Keytruda sales of $14.4 billion were more than double Opdivo's $7.0 billion. To this day, it is not clear if Keytruda is the better drug. But Merck designed a trial that was sufficient for success, whereas BMY overreached, with enduring consequences.

LET 'EM ROLL

Let 'Em Roll is a fun game to play and watch. Although the win rate was only 30% during seasons 47 and 48 (7 out of 23), the contestant often walked away with a few thousand dollars of prize money when failing to win. As such, *Let 'Em Roll* still ranked twenty-second in median earnings at $6,900, with the contestant garnering $2,450 in average earnings when failing to win the car or surpass $5,000 in cash.

Getting two or three rolls is key to success. When the contestant earned three rolls, he won four out of nine times, and when he got two rolls, he won two out of seven times. On the seven occasions in which the contestant got one roll, he never won the car, although he did earn $5,000 in cash once, which counted as a win. The average number of rolls was 2.1, meaning the contestant on average won only 1.1 out of 2 extra rolls. He hardly performed better than a coin toss in identifying whether the next prize was more or less expensive than the last. This is analogous to the poor performance of price ranking grocery items in *Hole in One*.

In analyzing *Let 'Em Roll* (illustrated in Figure 21.4), there are two key questions to ask.

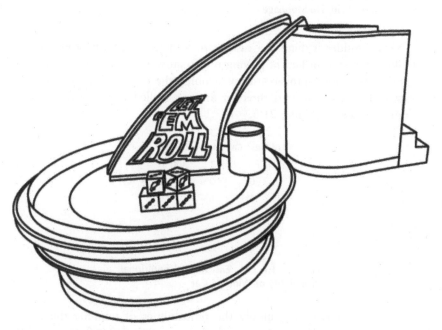

Figure 21.4 *Let 'Em Roll. Illustration courtesy of Omar Faruk*

1. What are the probabilities of winning the car given one, two, or three rolls?
2. Is there a cash amount that justifies stopping and walking away with the cash?

Given a cube with three cars and the typical $500, $1,000, and $1,500 cash prizes, the one-roll case is easy. The probability of winning the car is $1/2 \times 1/2 \times 1/2 \times 1/2 \times 1/2 = 1/32$, as 1/2 the faces on each cube are cash and 1/2 are cars. With probability 31/32, the contestant will walk away with cash averaging $2,500. Why? On average, 2.5 cubes will land on cash, and the average value will be $1,000 when the cube lands on a cash amount.

With two or three rolls, the odds increase, as seen in Figure 21.5. There are fifty-six different scenarios. I encourage readers to visit www.popculturemath. com for the detailed scenario analysis, but it rolls up to a 23.7% probability of winning the car with two rolls and a 51.3% probability with three rolls.

	Roll 1	Roll 2	Roll 3
Car Win Probability	3%	24%	51%
Expected Cash Win, no car	$2,581	$1,639	$1,363

Figure 21.5 *Let 'Em Roll*: Probability of Car Win and Expected Cash Win, No Stoppage

Now consider if there is a cash prize sufficient to justify stopping play and taking the cash in lieu of playing on. To answer this question, I first calculate the conditional probabilities of winning the car based on whether the contestant has zero, one, two, three, or four cars rolled with one or two rolls remaining, as seen in Figure 21.6.

# of Cars	Car Win Probability	
Roll 1	Roll 2	Roll 3
0	3%	24%
1	6%	32%
2	13%	42%
3	25%	56%
4	50%	75%

Figure 21.6 *Let 'Em Roll*—Conditional Probability of Winning Car

The next step is to multiply the probabilities of winning the car in Figure 21.6 by the price of the car, which I assume is $18,000, to gauge the amount of cash needed to justify stopping play. I assume the contestant is indifferent to a dollar in cash versus a dollar of expected car value. For example, 50% odds of winning the car would require $50\% \times \$18,000 = \$9,000$ cash to justify stopping, which exceeds the $7,500 maximum cash value (i.e., five $1,500 cards).

Consider if the contestant has one roll left, as shown in Figure 21.7. If zero cars are rolled, a cash value above $3,063 will justify stopping. There are 243 possible cash configurations across the five blocks, and 237, or 98%, will total more than $3,063. If the contestant has one roll remaining and one car rolled, a cash value of $3,125 will justify stopping, which will be achieved with 81% of rolls, representing sixty-six out of eighty-one possible cash configurations. Finally, if the contestant has two cars rolled and $4,000 or $4,500 in cash with one roll remaining, stopping is also in order. With two rolls remaining, we can run a similar analysis. Only if the contestant has zero cars rolled will stopping be in order and only if the cash value rolled is $6,000 or more, which will occur with 21% of rolls.

1 Roll Left – Upside from Rolling Again				Scenarios Justifying Stopping		
# of Cars Prior Roll	Exp Value of Car Win	Exp Value of Cash	Total Exp Value (EV)	# of Rolls > Total EV	Total Cash Combos	Probability Cash > EV
0	$563	$2,500	$3,063	237	243	98%
1	$1,125	$2,000	$3,125	66	81	81%
2	$2,250	$1,500	$3,750	4	27	15%
3	$4,500	$1,000	$5,500	0	9	0%
4	$9,000	$500	$9,500	0	3	0%

Figure 21.7 *Let 'Em Roll*—**Stopping Analysis**

During seasons 47 and 48, contestants never continued playing for the car when they would have been better off taking the cash. Two times contestants took the cash, and both were correct decisions. Specifically,

- On November 1, 2018, a contestant with two rolls got $4,000 and two cars on his first roll. The $4,000 cash value exceeded the $3,750 expected value of rolling again. He rightly stopped.
- On May 14, 2020, a contestant with two rolls got $4,000 and one car on his first roll. The $4,000 cash value exceeded the $3,125 expected value of rolling again, and he also rightly stopped.

All in all, contestant actions in *Let 'Em Roll* suggest a good intuitive grasp of probabilities. Interestingly, the 26.1% car win rate (6 out of 23) was almost spot-on with the 28.2% car win rate predicted when considering the probabilities of winning with one, two, and three rolls and the respective distribution of such rolls (7, 7, and 9 games out of 23). Specifically, $28.2\% = \left(\frac{7}{23} \times 3.1\%\right) + \left(\frac{7}{23} \times 23.7\%\right) + \left(\frac{9}{23} \times 51.3\%\right).$

If adjusting for the two instances in which the player correctly took the cash, the six out of twenty-one adjusted car win rate (28.6%) was spot-on with the 28.2% predicted.

Key Tip

Stop if your rolled cash value exceeds the expected value of rolling again for the car. This will usually mean stopping with zero or one car and one roll left (depending on the cash value rolled) and sometimes stopping with zero cars and two rolls left or two cars and one roll left.

MASTER KEY

Master Key is a simple game with success tied to the small item pricing, which determines whether the contestant gets zero, one, or two keys with which to try and win the car. If the contestant gets zero keys, he wins nothing. If he gets one key, he has a two in five chance (40%) of winning the car, as that key could be either the car key or the master key. If he gets two keys, there is a three in five chance that the first key is not a car-winning key. At that point, there are four keys left, and two of them are car-winning keys, creating a 50% chance of a car-winning key. So with two keys, the odds of winning are $\frac{2}{5} + \left(\frac{3}{5} \times \frac{1}{2} \right) = \frac{7}{10}$, or 70%.

With one key, if the contestant does not win the car, there is still only a one in five chance of going home empty-handed. With two keys, the contestant will never go home empty-handed, as there is only one blank key. Nor does the contestant risk the prizes he has banked when he tests the second key.

Given the expected win rates shown above (40% chance with one key, 70% chance with two keys), it may surprise readers that the win rate was 25% (4 out of 16) with only $6,500 in median earnings. The mediocre results turn out to be a function of chance. Contestants performed well in winning keys; in sixteen games they won a total of twenty keys: two keys seven times (7/16), one key six times (6/16), and zero keys three times (3/16). Given a maximum thirty-two keys that could have been won, the twenty keys won reflected a healthy 62.5% key win rate. Unfortunately, of the twenty keys won, six were blanks, six were for the third prize, and three were for the second prize. Only three keys were master keys, and only two keys were car keys.

Had contestants performed in line with the general odds based on the number of keys won, the car win rate would have been 46%, as follows:

$$\left(\frac{3}{16} \times 0\right) + \left(\frac{6}{16} \times \frac{2}{5}\right) + \left(\frac{7}{16} \times \frac{7}{10}\right) = \frac{12}{80} + \frac{49}{160} = \frac{73}{160} = 46\%$$

[3 × 0 keys] [6 × 1 key] [7 × 2 keys]

The 46% modeled win rate versus the 25% observed win rate would have shifted the median earnings from $6,500 to approximately $10,000, making *Master Key* one of the most lucrative pricing games.

There is one interesting heuristic in *Master Key*. Twenty-two out of thirty-two small items were captured by the last two digits in the three-digit number, and this held true in similar proportions during both seasons 47 and 48. For example, on January 14, 2020, the contestant was presented with a charcoal grill and the number string 576. He picked $57, and the right answer was $76. Is the overrepresentation of the last two digits due to chance? Unlikely. If there are 50/50 odds of each of the first two and last two numbers being correct, only 5% of the time would at least twenty-two out of thirty-two trials play out in favor of one set of numbers. So if the contestant is unsure as to whether to go with the first two numbers or the last two numbers, he should choose the last two digits.

Key Tip

Keys were won much more frequently when choosing the last two digits in the three-digit number. If undecided between the first two and last two numbers, choose the last two numbers.

RAT RACE

Rat Race is one of my favorite games. Contestants won eleven out of twenty times, for a 55% win rate and $10,950 in median earnings, which ranked fifth among all pricing games. Observed results were better than actual odds, as chance benefited contestants.

The math is simple. Five wind-up rats race and finish in random order. If a contestant prices the small items well, he can win three rats and have a three in five chance of winning the car. Winning two rats means a two in five chance of winning the car, and one rat a one in five chance. *Rat Race* is similar to *Master Key*, except that a contestant can win up to three rats, but only one rat (i.e., the winning rat) wins the car.

On average, contestants won 1.75 rats, or 58% of winnable rats, including three rats in four contests, two rats in nine contests, one rat in five contests, and no rats twice. The first rat and the second rat are harder to win, as the contestant must guess the price of a sub-$10 grocery item within $1 and the price of a sub-$100 small item within $10. For the third item, the contestant could have guessed $200 and won the third rat every single time, as the price was never less than $120 and never more than $300.

The contestant won the third rat on sixteen out of twenty tries, the second rat six times, and the first rat thirteen times. Poor performance on the second item also likely reflected unfamiliarity with the item in question, such as an aroma diffuser or tire inflator. Moreover, guesses for the second item were notoriously low. The average guess was $35 versus an average value of $57. That underbidding bias disappeared for the $100-plus item, in which the average guess was $208 versus an average price of $201, and for the sub-$10 item, in which the average guess was $4.13 and the average price was $4.48.[*]

Based on the number of rats won, one would have expected contestants to win the car 35% of the time:

$$\left[\left(4 \times \frac{3}{5}\right) + \left(9 \times \frac{2}{5}\right) + \left(5 \times \frac{1}{5}\right) + (2 \times 0)\right]/20 = \left(\frac{12}{5} + \frac{18}{5} + \frac{5}{5}\right)/20$$

$$= \frac{35}{100} = 35\%.$$

[4 × 3 rats] [9 × 2 rats] [5 × 1 rat] [2 × 0 rats]

The favorable 20 percentage point discrepancy (55% observed versus 35% modeled) reflected good luck, just as the negative 21 percentage point discrepancy in *Master Key* (25% observed versus 46% modeled) reflected bad luck. Notably, contestants won the car three out of four times with one rat in the race.

Key Tip

Guessing $200 for the $100-plus prize should allow a contestant to win the third rat without fail.

SPELLING BEE

Spelling Bee had a high win rate, with nine car wins out of nineteen iterations and $9,400 in median earnings. Had the contestant stopped with $5,000, that would have registered as a win, but that never transpired.[†]

---

[*]The $4.48 average ignores one egregiously high bid for the first item.

[†]We do not count as a win the Big Money Week show in which the player stopped at $15,000 (3 cards × $5,000 per card) when playing for a $100,000 Lexus sports car. Instead, we use $3,000 ($1,000 per card) for our median earnings calculation.

For *Spelling Bee*, the two key questions are:

1. What are the odds of winning *Spelling Bee* with three, four, or five cards?
2. When should the contestant stop and take the money versus playing for the car?

For the first question, the combinatorial math is rather complex; still, I encourage the reader to see if he or she can follow the three-card case here, and from there the four-card case and the five-card case.

Three-Card Case

There are thirty possible draws on the first card, twenty-nine on the second, and twenty-eight on the third, driving $30 \times 29 \times 28 = 24,360$ possible combinations.

How many of these combinations are winning combinations? Since there are eleven C's, eleven A's, and six R's, and the order of getting each letter does not matter, there are $11 \times 11 \times 6 \times 6 = 4,356$ possible winning combinations. Specifically, it does not matter if you pick C then R then A or C then A then R, so the six factor represents the six different ways of ordering the three letters. So 4,356 winning combinations divided by 24,360 total combinations results in a 17.9% probability of spelling "car."

Of course, the contestant can also win by drawing one of the car cards. What are the odds of this? Note that $\frac{28}{30} \times \frac{27}{29} \times \frac{26}{28}$ is the probability of not drawing a car card, as there are twenty-eight non-car cards on one's first draw, twenty-seven on the second, and twenty-six on the third. The probability of drawing a car card is thus $1 - \left[\left(\frac{28}{30} \right) \times \left(\frac{27}{29} \right) \times \left(\frac{26}{28} \right) \right] = 19.3\%$.

Since there is no overlap between winning by spelling "car" and winning by drawing a car card, the probability of winning with three cards is $17.9\% + 19.3\% = 37.2\%$.

Four-Card Case

With four cards, the math gets trickier. There are $30 \times 29 \times 28 \times 27 = 675,720$ possible combinations. Assuming no car card, a contestant wins if he gets two C's, one A, and one R, one C, two A's, and one R, or one C, one A, and two R's. Again, consider the total number of combinations and the number of winning combinations.

Winning combinations spelling "car" (and no car card) can be bucketed as follows:

- 2 C's, 1 A, 1 R = (11 C's × 10 C's × 11 A's × 6 R's) × 12 orderings
  = 87,120 winning combinations
- 1 C, 2 A's, 1 R = (11 C's × 11 A's × 10 A's × 6 R's) × 12 orderings
  = 87,120 winning combinations
- 1 C, 1 A, 2 R's = (11 C's × 11 A's × 6 R's × 5 R's) × 12 orderings
  = 43,560 winning combinations

The total number of combinations for spelling "car" in four draws is 217,800.

The last factor in the equations, 12, is the number of ways of ordering the four letters. For example, with two C's, one R, and one A, there are twelve possible orderings: *CCAR, CCRA, CACR, CARC, CRCA, CRAC, ACCR, ACRC, ARCC, RCCA, RCAC*, and *RACC*. The odds of spelling "car" is again based on winning combinations relative to total combinations: 217,800/657,720 = 33.1%.

The chance of a car card is calculated similarly to the three-card case—1 minus the probability of no car card:

$$1 - \left[\left(\frac{28}{30} \right) \times \left(\frac{27}{29} \right) \times \left(\frac{26}{28} \right) \times \left(\frac{25}{27} \right) \right] = 25.3\ \%.$$

The chance of winning with four cards is 25.3% + 33.1% = 58.4%.

Five-Card Case

The probability of spelling "car" without a car card is 42.5%, and the probability of winning with a car card is 31.0%, for a 73.5% chance of winning. This is shown in the Supporting Math appendix, along with a summary of the three-card, four-card, and five-card analysis.

All in all, the odds of winning *Spelling Bee* are quite good, explaining the high win rate. In the course of seasons 47 and 48, the contestant won one extra card for a total of three cards in nine out of nineteen games, two extra cards for a total of four cards in four games, and three extra cards for a total of five cards in six games. Based on the probabilities, the expected win rate, assuming contestants never chose cash, would have been:

$$\left(\frac{9}{19} \times 37.2\% \right) + \left(\frac{4}{19} \times 58.4\% \right) + \left(\frac{6}{19} \times 73.5\% \right) = 17.6\% + 12.3\%$$
$$+ 23.2\% = 53.1\%.$$

Interestingly, the presence of the two car cards means that it is never worthwhile to quit with one card left. With one's last card, the contestant has either a

2/26, 2/27, 2/28, or 2/29 chance of winning with a car card, depending on if one, two, three, or four cards have been revealed. Even a 2/29 chance of drawing a car card and winning a modest $18,000 car has a $\frac{2}{29} \times \$18,000 = \$1,241$ expected value, superior to $1,000 in cash. If the contestant has two cards in *Spelling Bee*, by virtue of failing to win an additional card, he can win only with a car card, which occurs with $1 - \left(\frac{28}{30}\right) \times \left(\frac{27}{29}\right) = 13.1\%$ probability.

A 13.1% chance of winning an $18,000 car equates to $2,358 in expected value, which is greater than $2,000 cash. And if the contestant has a chance of spelling "car" with one or two cards left, his odds of winning will be higher than those shown here, which are tied solely to getting a car card. Of course, the contestant may prefer a lesser amount of cash to the expected value of winning the car if he (1) prefers cash, (2) is risk averse, or (3) does not want the car offered.

One can quickly verify that a contestant should *never* take the cash with three, four, or five cards. A 37.2% chance of winning an $18,000 car with three cards has an expected value of $6,700, significantly more than $3,000, and the expected value of a car win with four or five cards far exceeds the $4,000 or $5,000 in cash.

Seemingly contrary to their best interests, contestants stopped with cash five out of nineteen times, including once with four cards, twice with three cards, and once with two of three cards left. They would have won the car three of those five times, including a $100,000 Lexus sports car on February 19, 2020, during Dream Car Week. Here the contestant accumulated $15,000 across three cards in special $5,000 increments and stopped before any cards were played. He would have won the Lexus with a *C*, an *A*, and an *R* card had he played on. Understandable given the cash involved, but not optimal.

Had contestants *never* walked with cash, they would have won a car twelve out of nineteen times (63%) versus the nine out of fourteen times (64%) observed. Similar win rate but more wins. Contestants won the car two out of nine times with three cards, two out of four times with four cards, and five out of six times with five cards.

Contestants averaged 1.85 additional cards per game (i.e., 3.85 total cards), a 62% card win rate. They could have averaged 2.4 additional cards per game (i.e., 4.4 total cards) by choosing $30 for the first prize, $40 for the second prize, and $35 for the third prize, even without knowing what the prizes were. Contestants might have won even more cards if they started with $30/$40/$35 and then tweaked their guesses up or down as appropriate. The small items in *Spelling Bee* did not have quite as narrow of a range as in *Cliffhanger* but narrow enough to take advantage of.

Key Tip

Maximize your expected value by always playing for the car, even when the odds seem low. Additional picks may be won with greater frequency by guessing $30/$40/$35 for the first/second/third small items and deviating by $5 if the item seems more or less pricey.

Appendix: Supporting Math

SHOWCASE SHOWDOWN—CONTINUOUS SOLUTION

Our analysis of the Showcase Showdown in chapter 4 relied on a Monte Carlo simulation, which is a useful and robust technique. However, we can also use calculus to solve a continuous wheel solution. From there, we can take the optimal stopping point for Spinner 2 and Spinner 3 and consider the $0.05 interval just above and just below the discrete solution, noting that tie spins are a real-life consideration that do not come into play for a continuous wheel (since numbers chosen randomly out of an infinite set will never be exactly equal).

Consider Spinner 1. If Spinner 1 spins x, he has a probability x^2 of prevailing over Spinner 2. That is because his spin of x will beat Spinner 2's first spin of y with probability x and beat Spinner 2's second spin of y_2 with probability x, for a multiplicative success probability of x^2. For example, if Spinner 1 spins $0.65 and stays, he has a 65% chance of being ahead of Spinner 2 after Spinner 2's first try and a 65% chance of being ahead after Spinner 2's second try, for a $0.65 \times 0.65 = 0.42 = 42\%$ expected win rate. But if he spins again, he goes over 65% of the time.

Similarly, if Spinner 1 is still in the lead after facing Spinner 2—because Spinner 2 comes up short or goes over on his second spin—his probability of prevailing over Spinner 3 is also x^2. Spinner 1's probability of prevailing over both Spinner 2 and Spinner 3 is therefore $x^2 \times x^2 = x^4$. This is Spinner 1's chance of victory if he spins only once.

What is Spinner 1's probability of success if he spins again? Consider his two-spin total $v = x + x_2$. For every two-spin total value $v < 1$, his probability of victory is v^4. But we need to sum all values that do not take $v = x + x_2$

over 1. To do that we integrate the second spin v from x to 1, per $\int_x^1 v^4 \, dv$, which gives us $[\frac{1}{5} - \frac{x^5}{5}]$.

We set $x^4 = \frac{1}{5} - \frac{x^5}{5}$ to find the number x between 0 and 1 that creates an equal probability of victory for Spinner 1 when spinning once (probability x^4) and spinning twice (probability $[\frac{1}{5} - \frac{x^5}{5}]$). For initial spins above this value, $\frac{1}{5} - \frac{x^5}{5} < x^4$ means there is a higher probability of victory in not spinning again. For spins below this value, $x^4 < \frac{1}{5} - \frac{x^5}{5}$ means there is a higher probability of victory in spinning again. Since there is no simple discrete solution for a polynomial equation of power 5, Figure A.1 compares the values of x^4 and $\frac{1}{5} - \frac{x^5}{5}$ between 0 and 1. Figure A.1 suggests an indifference point of 0.65.

What about Spinner 2? His first order of business is to try and beat Spinner 1. If he does not do so on his first spin, he spins again. But what

Spinner 1 Probability of Victory

1st Spin	Shifted 1st Spin	P(one spin) =x^4	P(two spins) =1/5 - x^5/5	P(one spin) - P(two spins)	
0.41	0.435	2.8%	19.8%	-16.9%	
0.42	0.445	3.1%	19.7%	-16.6%	
0.43	0.455	3.4%	19.7%	-16.3%	
0.44	0.465	3.7%	19.7%	-15.9%	
0.45	0.475	4.1%	19.6%	-15.5%	
0.46	0.485	4.5%	19.6%	-15.1%	
0.47	0.495	4.9%	19.5%	-14.7%	
0.48	0.505	5.3%	19.5%	-14.2%	
0.49	0.515	5.8%	19.4%	-13.7%	
0.50	0.525	6.3%	19.4%	-13.1%	
0.51	0.535	6.8%	19.3%	-12.5%	
0.52	0.545	7.3%	19.2%	-11.9%	
0.53	0.555	7.9%	19.2%	-11.3%	
0.54	0.565	8.5%	19.1%	-10.6%	
0.55	0.575	9.2%	19.0%	-9.8%	
0.56	0.585	9.8%	18.9%	-9.1%	
0.57	0.595	10.6%	18.8%	-8.2%	
0.58	0.605	11.3%	18.7%	-7.4%	
0.59	0.615	12.1%	18.6%	-6.5%	
0.60	0.625	13.0%	18.4%	-5.5%	
0.61	0.635	13.8%	18.3%	-4.5%	Spin
0.62	0.645	14.8%	18.2%	-3.4%	Again
0.63	0.655	15.8%	18.0%	-2.3%	
0.64	0.665	16.8%	17.9%	-1.1%	
0.65	0.675	17.9%	17.7%	0.2%	
0.66	0.685	19.0%	17.5%	1.5%	
0.67	0.695	20.2%	17.3%	2.9%	Do not
0.68	0.705	21.4%	17.1%	4.3%	Spin
0.69	0.715	22.7%	16.9%	5.8%	Again
0.70	0.725	24.0%	16.6%	7.4%	
0.71	0.735	25.4%	16.4%	9.0%	
0.72	0.745	26.9%	16.1%	10.7%	
0.73	0.755	28.4%	15.9%	12.5%	
0.74	0.765	30.0%	15.6%	14.4%	
0.75	0.775	31.6%	15.3%	16.4%	
0.76	0.785	33.4%	14.9%	18.4%	
0.77	0.795	35.2%	14.6%	20.6%	
0.78	0.805	37.0%	14.2%	22.8%	
0.79	0.815	39.0%	13.8%	25.1%	
0.80	0.825	41.0%	13.4%	27.5%	

Spinner 2 Probability of Victory

1st Spin	Shifted 1st Spin	P(one spin) =x^2	P(two spins) =1/3 - x^3/3	P(one spin) - P(two spins)	
0.41	0.435	16.8%	31.0%	-14.2%	
0.42	0.445	17.6%	30.9%	-13.2%	
0.43	0.455	18.5%	30.7%	-12.2%	
0.44	0.465	19.4%	30.5%	-11.1%	
0.45	0.475	20.3%	30.3%	-10.0%	
0.46	0.485	21.2%	30.1%	-8.9%	
0.47	0.495	22.1%	29.9%	-7.8%	
0.48	0.505	23.0%	29.6%	-6.6%	
0.49	0.515	24.0%	29.4%	-5.4%	
0.50	0.525	25.0%	29.2%	-4.2%	
0.51	0.535	26.0%	28.9%	-2.9%	
0.52	0.545	27.0%	28.6%	-1.6%	
0.53	0.555	28.1%	28.4%	-0.3%	
0.54	0.565	29.2%	28.1%	1.1%	
0.55	0.575	30.3%	27.8%	2.5%	
0.56	0.585	31.4%	27.5%	3.9%	
0.57	0.595	32.5%	27.2%	5.3%	
0.58	0.605	33.6%	26.8%	6.8%	
0.59	0.615	34.8%	26.5%	8.3%	
0.60	0.625	36.0%	26.1%	9.9%	
0.61	0.635	37.2%	25.8%	11.4%	Spin
0.62	0.645	38.4%	25.4%	13.1%	Again
0.63	0.655	39.7%	25.0%	14.7%	
0.64	0.665	41.0%	24.6%	16.4%	
0.65	0.675	42.3%	24.2%	18.1%	
0.66	0.685	43.6%	23.8%	19.8%	Do not
0.67	0.695	44.9%	23.3%	21.6%	Spin
0.68	0.705	46.2%	22.9%	23.4%	Again
0.69	0.715	47.6%	22.4%	25.2%	
0.70	0.725	49.0%	21.9%	27.1%	
0.71	0.735	50.4%	21.4%	29.0%	
0.72	0.745	51.8%	20.9%	30.9%	
0.73	0.755	53.3%	20.4%	32.9%	
0.74	0.765	54.8%	19.8%	34.9%	
0.75	0.775	56.3%	19.3%	37.0%	
0.76	0.785	57.8%	18.7%	39.1%	
0.77	0.795	59.3%	18.1%	41.2%	
0.78	0.805	60.8%	17.5%	43.3%	
0.79	0.815	62.4%	16.9%	45.5%	
0.80	0.825	64.0%	16.3%	47.7%	

Figure A.1 "Continuous Wheel"—Probability of Victory

if Spinner 1 has gone over or Spinner 2 has surpassed Spinner 1 on his first spin? When should he spin again, and when should he stay? Assuming Spinner 1 is no longer a consideration, Spinner 2 wins with probability y^2 by staying after his first spin and $\int_y^1 z^2 \, dz$ by spinning again, where $z = y + y_2$. Spinner 2 has to beat only two tries by one player versus Spinner 1 beating two tries by each of two players, so we use only the squaring function for Spinner 2's first spin y (versus taking Spinner 1's first spin x to the fourth power). Setting $y^2 = \int_y^1 z^2 \, dz = \frac{1}{3} - \frac{y^3}{3}$ allows us to solve for threshold level y at which the odds of victory are balanced between spinning again and staying. Although cubic equations have a general solution, it is rather messy, and we again compare probabilities in Figure A.1, which suggests an indifference point of 0.53.

Now I make one adjustment. The continuous analogue to a wheel with $0.05 cent increments starting at $0.05 and increasing to $1.00 is not a continuous wheel from 0 to 1. Rather, it is a continuous wheel from $0.025 to $1.025. In effect, spinning between 0.025 and 0.075 matches with 0.05, whereas spinning from 0.975 to 1.025 matches with 1.00, and the same would apply for every $0.05 interval in between. This means we need to shift up the solutions for Spinner 1 and Spinner 2 by 2.5 cents to correspond to the Showcase Showdown.

Consequently, Spinner 2's ideal stopping point in the continuous game with a wheel ranging from $0.025 to $1.025 shifts to $0.555 from $0.53 cents, and Spinner 1's stopping point shifts to $0.675 from $0.65. At this point we are reasonably sure that $0.65 or $0.70 is the appropriate stopping point for Spinner 1 and reasonably sure that $0.55 or $0.60 is the appropriate stopping point for Spinner 2, as we expect ties to influence the continuous wheel solution only on the margin.

The bias downward in Spinner 2's ideal stopping point (versus considering only $0.60) relates to the effect of ties. The value that is the inflection point between spinning again and staying is the value that leads to the lowest probability of success. The odds of success for Spinner 2 at 55.5 cents in the continuous solution are slightly more than 28%. The odds of success for Spinner 1 at $0.675 in the continuous game are slightly less than 18%. Imagine as Spinner 2 that you spin again at $0.55. You go over with $9/20 = 45\%$ probability, and you are not guaranteed victory if you avoid going over. But if you stop at $0.55 cents and Spinner 3 ties you on his first spin, he will prefer a spinoff to trying to beat you on his second spin. Generalizing this example, ties always slightly increase the relative probability associated with stopping at a lower value versus spinning again, provided that value is at least $0.50. Said another way, a tie is a tie at $0.50 or more, regardless of whether one ties at $0.55 or a higher number; for example, $0.90.

The intuitive conclusion from the analysis of ties suggests that Spinner 1 should stop at $0.65 and higher (not $0.70 and higher) and Spinner 2 at $0.55 and higher (not $0.60 and higher). We could confirm this intuitive result by running a small-scale Monte Carlo simulation comparing the relative win rates. This is not necessary given the comprehensive Monte Carlo simulations performed in chapter 6 to solve the Showcase Showdown. Note that Spinner 1 ought to stop at $0.70, not $0.65, as ties benefit Spinner 1 less.

Finally, and of interest, we can sum up Spinner 1's probability of success in the continuous game. Since each outcome 0.01 through 1.00 is equally likely on the first spin, we average the probability of victory across each first spin, whether that probability is maximized by spinning again or staying after the first spin. We calculate an approximate 30.4% chance of success for Spinner 1 versus Spinner 2 and Spinner 3. The 30.4% chance of success for Spinner 1 compares to the 31.5% success in a Monte Carlo simulation, the difference in part relating to the benefit of ties for Spinner 1, as captured in the Monte Carlo simulation.

We can similarly calculate a 45.4% chance of success for Spinner 2 versus Spinner 3 provided that Spinner 2 has "neutralized" Spinner 1 by virtue of Spinner 1 going over or being surpassed by Spinner 2 on Spinner 2's first spin. This analysis, however, does not lead to an easy approximation for Spinner 2's overall win rate, as we would need to calculate what percentage of the time Spinner 1 is "neutralized," which is computationally complex.

PUNCH-A-BUNCH: EXPECTED VALUE ANALYSIS

So how do we calculate the expected win value with two punches? We average the expected win value on a winning first punch and the expected win value on a second punch if the contestant plays on. A winning first punch of $2,500 or more has an average value of $85,000/15 = $5,670, reflecting $85,000 value in aggregate across the fifteen cards with a value of at least $2,500. The other 70% of the time, the contestant wins $1,000 or less and is better off looking at the second punch. A sub-$2,500 first punch has an expected value of $18,000/35 = $515, reflecting $18,000 of value in aggregate across thirty-five cards with a value of no more than $1,000. The remaining board now has a slightly higher expected value of $($103,000 − 515)/49 = $2,090$, subtracting the $515 average value of the returned card. The expected win with two punches is therefore $(30\% \times \$5,670) + (70\% \times \$2,090) = \$3,160$.

1st Punch			2nd Punch			3rd Punch		
	Prob	Exp Win						
>= $5,000	14%	$ 9,286		Prob	Exp Win			
< $5,000	86%	$ 884 ->	>=$2,500	31%	$ 5,667		Prob	Exp Win
			< $2,500	69%	$ 503 ->	3rd Punch	100%	$ 2,117

Expected Value = (14% * $9,286) + (86% * 31% * $5,667) + (86% * 69% * $2,117)   = $ 4,055

1st Punch			2nd Punch			3rd Punch			4th Punch		
	Prob	Exp Win									
>= $5,000	14%	$ 9,286		Prob	Exp Win						
< $5,000	86%	$ 884 ->	>= $5,000	14%	$ 9,286		Prob	Exp Win			
			< $5,000	86%	$ 884 ->	>= $2,500	31%	$ 5,667		Prob	Exp Win
						< $2,500	69%	$ 492 ->	4th Punch	100%	$ 2,143

Expected Value = (14% * $9,286) + (86% * 14% * $9,286) + (86% * 86% * 31% * $5,667) + (86% * 86% * 69% * $2,143)   = $ 4,832

Figure A.2 Expected Value of Punch-a-Bunch with Three and Four Punches

With three punches, the contestant's strategy changes. Since two punches yields an expected win of $3,160, the contestant on his first punch should no longer stop with a $2,500 first punch, because he now has two additional punches, each with an expected win of roughly $3,160. Instead, he should stop at $5,000 and more on the first punch and $2,500 and more on the second punch. With four punches, the contestant gets more shots at a big prize. The math is such that he should stop at $5,000 or more on his first two punches and $2,500 or more on his third punch. Figure A.2 shows the detailed math. Notably, a third punch increases the expected win value to more than $4,000 and a fourth punch to more than $4,800. Accordingly, the contestant should stop at $5,000 or more with two or more punches left and $2,500 with one punch remaining.

SPELLING BEE

Five-card scenario: The probability of spelling "car" without a car card is 42.5%, and the probability of winning with a car card is 31.0%, for a 73.5% chance of winning (Figure A.3).

Four-card scenario: The probability of spelling "car" without a car card is 33.1%, and the probability of winning with a car card is 25.3%, for a 58.4% chance of winning.

Three-card scenario: The probability of spelling "car" without a car card is 17.9%, and the probability of winning with a car card is 19.3%, for a 37.2% chance of winning.

5-Card Winning Combinations

C	A	R	1st C	2nd C	3rd C	1st A	2nd A	3rd A	1st R	2nd R	3rd R	Number of Orderings	Winning Combinations
3	1	1	11	10	9	11			6			20	1,306,800
2	2	1	11	10		11	10		6			30	2,178,000
2	1	2	11	10		11			6	5		30	1,089,000
1	1	3	11			11			6	5	4	20	290,400
1	3	1	11			11	10	9	6			20	1,306,800
1	2	2	11			11	10		6	5		30	1,089,000
													7,260,000

# Combinations	$\dfrac{30!}{(30-5)!}$	17,100,720
P (spell CAR & no CAR card):		42.5%
P (drawing CAR Card)	->	31.0%
= 1-[(28/30)*(27/29)*(26/28)*(25/27)*(24/26)]		
Probability of Success		73.5%

4-Card Winning Combinations

C	A	R	1st C	2nd C	3rd C	1st A	2nd A	3rd A	1st R	2nd R	3rd R	Number of Orderings	Winning Combinations
2	1	1	11	10		11			6			12	87,120
1	2	1	11			11	10		6			12	87,120
1	1	2	11			11			6	5		12	43,560
													217,800

# Combinations	$\dfrac{30!}{(30-4)!}$	657,720
P (spell CAR & no CAR card):		33.1%
P (drawing CAR Card)	->	25.3%
1-[(28/30)*(27/29)*(26/28)*(25/27)]		
Probability of Success		58.4%

3-Card Winning Combinations

C	A	R	1st C	2nd C	3rd C	1st a	2nd A	3rd A	1st R	2nd R	3rd R	Number of Orderings	Winning Combinations
1	1	1	11			11			6			6	4,356

# Combinations	$\dfrac{30!}{(30-3)!}$	24,360
P (spell CAR & no CAR card):		17.9%
P (drawing CAR Card)	->	19.3%
1-[(28/30)*(27/29)*(26/28)]		
Probability of Success		37.2%

Figure A.3 Spelling Bee—Probability of Winning Car